Eugenijus Gaubas
Tomas Ceponis
Juozas-Vidmantis Vaitkus

Pulsed capacitance technique for evaluation of barrier structures

AF138225

Eugenijus Gaubas
Tomas Ceponis
Juozas-Vidmantis Vaitkus

Pulsed capacitance technique for evaluation of barrier structures

LAP LAMBERT Academic Publishing

Impressum / Imprint

Bibliografische Information der Deutschen Nationalbibliothek: Die Deutsche Nationalbibliothek verzeichnet diese Publikation in der Deutschen Nationalbibliografie; detaillierte bibliografische Daten sind im Internet über http://dnb.d-nb.de abrufbar.
Alle in diesem Buch genannten Marken und Produktnamen unterliegen warenzeichen-, marken- oder patentrechtlichem Schutz bzw. sind Warenzeichen oder eingetragene Warenzeichen der jeweiligen Inhaber. Die Wiedergabe von Marken, Produktnamen, Gebrauchsnamen, Handelsnamen, Warenbezeichnungen u.s.w. in diesem Werk berechtigt auch ohne besondere Kennzeichnung nicht zu der Annahme, dass solche Namen im Sinne der Warenzeichen- und Markenschutzgesetzgebung als frei zu betrachten wären und daher von jedermann benutzt werden dürften.

Bibliographic information published by the Deutsche Nationalbibliothek: The Deutsche Nationalbibliothek lists this publication in the Deutsche Nationalbibliografie; detailed bibliographic data are available in the Internet at http://dnb.d-nb.de.
Any brand names and product names mentioned in this book are subject to trademark, brand or patent protection and are trademarks or registered trademarks of their respective holders. The use of brand names, product names, common names, trade names, product descriptions etc. even without a particular marking in this works is in no way to be construed to mean that such names may be regarded as unrestricted in respect of trademark and brand protection legislation and could thus be used by anyone.

Coverbild / Cover image: www.ingimage.com

Verlag / Publisher:
LAP LAMBERT Academic Publishing
ist ein Imprint der / is a trademark of
OmniScriptum GmbH & Co. KG
Heinrich-Böcking-Str. 6-8, 66121 Saarbrücken, Deutschland / Germany
Email: info@lap-publishing.com

Herstellung: siehe letzte Seite /
Printed at: see last page
ISBN: 978-3-659-50518-8

Content

Introduction

Potential barrier (U_b) in junction structure is one of the most important parameter in description and analysis of the operational characteristics of semiconductor devices. Changes of the potential barrier under applied external voltage source depend on voltage polarity, on voltage value, on rate of voltage changes (frequency and frequency spectrum) and on capabilities of electrodes (ohmic, blocking, injecting). Commonly, characteristics of current changes under applied voltage (the so called I-V technique in DC regime) and of capacitance dependent on voltage (the so called C-V) are exploited to evaluate barrier parameters.

The most general consideration is based on Maxwell's equations in matter. To describe the transient electrical characteristics in matter, the static and dynamic approaches exist [1]. The static theories are devoted to description of a drift velocity versus electric filed characteristic. The common feature of static and quasi-steady-state approaches is a consideration of only the static and uniform electric fields. The most general dynamic theories deal with the charge transport properties under dynamic conditions when electric field varies in time and space. However, in most of text books on semiconductor devices the paradigm of drift current description is based on consideration of the first Poisson's equation and drift-diffusion current model, which actually can be ascribed to quasi-steady state approaches. The first Poisson's equation calibrates the charge in the system. Charge balancing through the changes of image charge leads to an appearance of electric fields varied in time and space. Thus, direct insertion of quasi-static electric fields, described by the first Poisson's equation, into expression of the drift and diffusion currents leaves a system non-calibrated respective to a potential. Addition of the continuity equation, to describe the current changes in time, does not improve precision of description. In continuity equation, description of current and of carrier trapping and recombination is tacitly calibrated by charge balance. This balance is also determined by a precision of the potential calibration. The non-calibrated potential determines precision of the approximations derived from drift-diffusion approach. An example can be a consideration of recombination models and continuity equation, where necessary to precisely include the balance of charged carrier flows, of group velocity and of electric fields [2, 3]. The main inconsistency of such models appears due to non-calibrated potential in field equations, i.e. by ignoring of the image charge. The main relation of electrostatics requires the unambiguous correlation between the charge and potential. This correlation is calibrated by the second Poisson's integral including impact of the image charge, which is actually provided by external source. The system of field description becomes concerted if the charges correlated with calibrated potential are accounted for within the first Poisson's equation. For non-calibrated consideration of the drift-diffusion approximation, this approach is only valid for partial description, - for instance, drift-diffusion can be applied for description of current in the transitional layers nearby the field discontinuity interfaces (e.g. electrodes) after rapid transitional processes of field discontinuity drift are finished or balanced. Calibration of the potential (of acting voltage U) is directly related with characteristic lengths (d) under consideration. In this monograph, calibration of the charge and potential is always controlled and discussed.

It has been shown [4, 5] that analysis of the capacitance changes in time is equivalent to a consideration of current transients. It is obvious that measurements of the capacitance characteristics can only be implemented by using variable voltage source. There are several techniques for C-V measurements [6].

One of them is a traditional technique using a harmonic test signal [6] $U(t)=U_{\sim}\exp(i\omega t)$ of fixed frequency ω. For junction structures, an additional requirement of $eU_{\sim}/kT<<1$ and $U_{\sim}<<U_b$ should be kept to avoid the generation of harmonics within a non-linear junction. The test signal technique (TST) is based on measurements of phase shift between a reference test voltage signal and a current, which appears due to barrier capacitance charging. However, these TST measurements provide the correct barrier capacitance response if a displacement current dominates and a frequency of the test signal is not too high $\omega<1/\tau_{NDef}$ relative to a dielectric relaxation time $\tau_{NDef}=\varepsilon\varepsilon_0/e\mu n_0$ within a material of dielectric permittivity ε with equilibrium carrier density n_0 and carrier mobility μ. Here, additional symbols denote: e is the elementary charge and ε_0 is the vacuum permittivity. In the range of low frequencies for TST technique, additional limitation appears due to deep traps in material. The generation current component appears due to thermal emission of trapped carriers. The generation current is able to completely disturb and hide the displacement current in TST if density of traps is close or even exceeds the density $n_0=N_{Def}=N_D-N_A$ of shallow dopants. N_{Def} determines a sufficiently rapid response of displacement current with τ_{NDef}. The generation current leads to a rapid (in terms of low frequency) dissipation component, which diminishes the quality factor [6] of a LRC-meter to the un-acceptable values. In the range of low frequencies $1/\tau_{gen}\leq\omega<<1/\tau_{NDef}$, the temperature changes of generation current can be applied for spectroscopy of deep levels (the so called C-DLTS [6, 7] technique).

However, validity of the TST method and instrumentation for usage in DLTS is only held if trap density N_T is sufficiently small in respect to N_{Def},- in the case of $N_T > N_{Def}$, a potential barrier is completely determined by N_T due to requirement of charge conservation. Then, barrier charging becomes rather slow: $\tau_{gen} \sim \exp(E_T/kT)$ for deep traps ($E_T >> kT$) with activation energy E_T relative to a thermal energy kT. These rather slow changes of barrier charging current can then be exceeded by Shockley-Ramo's [4, 5, 8] current changes due to a drift of the thermally emitted carriers. Analysis of drift current induced by the thermally injected charge can also be applied for spectroscopy of deep traps even if $N_T > N_{Def}$, - the so called I-DLTS technique. However, separation of the barrier charging (displacement) and of convection (drift) current components is extremely complicated in I-DLTS measurements using common techniques. An alternative technique to I-DLTS is the measurement of drift current due to optically injected charge, - the so called O-I-DLTS technique. There, separation of the current components attributed to the thermally injected charge and to that of trapping of the optically injected carriers is again complicated. Therefore, development of new barrier evaluation techniques is desirable. One of the aims of this work has been addressed to creation of a new technique, which would allow the separating of the barrier charging and generation current components.

Alternatively to the TST measurements, several pulsed techniques are exploited to characterize barriers. One of them, the reverse recovery (RR) of a junction technique [2, 6] is employed to examine the rate of extraction of the injected (during turn-on state of junction) minority carrier charge. Actually, this technique is based on examination of a transient of the recovery of diffusion (storage) capacitance. A switching pulse of the external current (voltage) is there applied to commute a junction from the forward voltage regime to a reverse one. Commonly, either a square-wave or linearly decreasing current pulse is employed [2, 6]. Application of the square-wave pulse leads to a rather considerable inductance current when reverse current rises sharply to a peak value under the reverse voltage shot. There, barrier is switched from the storage capacitance to the barrier capacitance regime by a rather long stage of the invariable current, which stability is supported by inductance current. The inductance current is determined by the carrier recombination and diffusion processes within a junction base region. The technique of the linearly increasing voltage (LIV) pulse of the reverse polarity enables ones to suppress sufficiently the inductance current. This LIV technique is preferential relative to the square-wave pulse regime in order to examine the softness (symmetry) of the switching characteristic [2]. These characteristics are of paramount importance in evaluation of junction operation within the switching regimes of power transducers.

Actually, a circuit containing a pulse generator and an RC chain consisting of a load resistor and a device under test (DUT) connected in series is exploited in routine measurements using pulsed techniques. In more general view, the capacitance component determines the final stage of a pulsed response within the transitional recharging processes in this RC chain. Simple Fourier analysis hints that square-wave pulse of a generator is transformed into the triangle response pulse for RC chain. In contrary, the LIV pulse of a generator is transformed into the square-wave response pulse for RC chain.

A sharp voltage step of the square-wave generator induces inevitably the Ramo's current within a regime of $\tau_{Mq} \ll \tau_{NDef}$ [5],- with a short time τ_{Mq} of the field discontinuity motion and induction current pulse due to charge extraction (drift). As a consequence, the initial shot of a pulse differentiation appears within a response pulse. The latter differentiation shot is also transformed by the voltage drop on a load resistor. These current components significantly complicate analysis of the barrier capacitance changes under applied square-wave voltage pulse.

As mentioned above, a pulse of the square-wave response should be observed under LIV pulse, if a component of the capacitance charging current dominates in DUT. Then, the charge extraction current follows the LIV generator signal, where Ramo's current regime of $\tau_{Mq} \geq \tau_{NDef}$ [5] can be implemented. This enables ones to directly examine the junction barrier and material. Therefore, barrier evaluation using LIV pulses, - the BELIV technique, is preferential for examination of junction structures. This monograph is addressed to description of the peculiarities of the BELIV technique. As it will be shown in the text, the BELIV technique has also the additional merit, - it allows separating in time the components of the barrier charging and the generation current.

Description of the operational characteristics of devices is based on different approaches (such as the drift-diffusion, hydrodynamics, and electrostatics [1, 4, 5, 7, 8]) depending on time scale of considered phenomena and device functioning regimes. These description approaches lead to the solution of either Cauchy or boundary tasks for the system of differential equations. However, in traditional text-books on semiconductor device physics [2, 9, 10], a clear formulation of mathematical tasks and discussion concerning validity of the obtained solutions is often ignored. For instance, the drift-diffusion model is only valid for analysis of the quasi-steady-state regime. Actually, it is a Cauchy task, which can be applied for analysis of an interface within a sample of infinite dimensions. This $d \rightarrow \infty$ immediately leads to ignoring of the role of external voltage source ($U \rightarrow 0$). Then, the diffusion of carriers serves as a source for charge separation and appearance of the electric field within interface (the transitional layer of thickness λ [7]) between the depletion and the electrically neutral region (ENR). Thus, drift-diffusion model can only be applied if a charge changes on electrodes (due to action of the external voltage source in charge balancing) might be ignored. It will be shown that the BELIV pulsed technique with LIV pulse τ_{LIV} durations of $\tau_{LIV} > \tau_{Mq} \geq \tau_{NDef}$ can be employed for the reliable and simple analysis of parameters of the junction barrier. On the other hand, drift-diffusion model can be applied for analysis of charge balance nearby an ohmic electrode which determines a quasi-steady-state current within external circuit. For description of charge extraction within time intervals $\tau_{LIV} \cong \tau_{Mq}$ $\cong \tau_{NDef}$ only the electrostatic approach is acceptable. There, the Ramo's current components should directly be considered with assumption of an infinite rate of the electrostatic induction, however, by taking into account conservation of charge, of charge momentum and of electrostatic energy. The Ramo's regime is held if $\tau_{Mq} = \tau_{TOF}$ [5], i.e. a duration of an action of the external voltage source (τ_{TOF}) should be equal to the reaction time τ_{Mq} of a moving field discontinuity. Actually, such processes are inherent for extremely short LIV pulses. Then, only very short processes of charge trapping or generation from shallow trap levels can take place. Analysis of these Shockley- Ramo's currents [4, 8] in junction structures can be found in [5], however, in common text-books on semiconductor device physics this consideration is often ignored. The relevant description approaches will be discussed in this monograph when necessary.

In most of applications of the BELIV technique, described in this monograph, rather long LIV pulses $\tau_{LIV} >> \tau_{NDef} > \tau_{Mq} = \tau_{TOF}$ have been employed. Thus, it is commonly assumed that transitional layer, which thickness $\lambda \cong L_D$ can be estimated by Debye screening length L_D, is stabilized. In such an approach, other current components (diffusion, generation, recombination) are evaluated by using the quasi-steady-state approaches.

The aim of development of the BELIV technique was to overcome limitations of the test signal (TST) and pulsed techniques. BELIV technique is free of limitation inherent for DLTS techniques, appeared due to sample geometrical capacitance. In DLTS, the constant capacitance of a device should be compensated in measurements of capacitance transients. For BELIV technique, the response signal is even increased for large dimension samples. Therefore, BELIV technique can widely be applied for spectroscopy of deep traps in rather heavily doped materials using the enhanced dimension samples. Such the structures are inherent for solar cells. Possibility to separate the components of the barrier capacitance charging and the generation currents in time scale of the BELIV transients enables ones to examine spectra of deep traps of elevated densities with $N_T > N_{Def}$.

Another preferential feature of BELIV technique is imaging of the inherent shapes of BELIV transients,- i) the transient has a triangle shape (repeating pulse generator) if interface is ohmic; ii) the response is square-wave shape if barrier is completely depleted; iii) the hyperbolic descendent shape is inherent for barrier capacitance charging curve; iv) the transient contains a minimum between the descending and ascending branches of a transient if generation current competes with barrier capacitance charging one in the time scale of LIV pulse; v) prevailing of generation current is indicated by a saturated shape of the BELIV transient. By varying LIV pulse durations, transforms of BELIV transient shapes can be analyzed. This enables ones to measure separately the parameters of different processes.

Also, the multilayered structures can be scanned by using probe positioned on a cross-sectional boundary of a structure,- interfaces can directly be highlighted by the changes of a transient shape, while doping parameters are extracted by using BELIV transients ascribed to a definite layer. Analysis of transients and their changes under steady-state or spectrally resolved pulsed bias illumination allows examination of barrier and deep trap parameters ascribed to a definite layer.

The configurational diagram of a definite trap is a portrait of this electrically active defect. To make the configurational diagram, the photo-ionization/photo-neutralization spectra should be examined together with thermal emission spectra, including lattice relaxation processes. Analysis of the time evolution of BELIV transients using spectrally resolved excitation pulses enables ones to resolve the current components determined by photo- and thermal- excitation. Thereby, BELIV technique can be employed for complete spectroscopy of deep traps in material used for junction fabrication.

A rather simple circuit for the BELIV technique implementation is suitable for the on-line (in situ) control of barrier evolution during irradiation by different species of accelerated particles. These measurements can be performed within a harsh environment of accelerated particle beams.

These above mentioned applications of the BELIV technique are demonstrated in this monograph. To clarify the details of the BELIV technique, important for

8

applications, the measurement circuitries and their impact are initially described in this book. Then, the detail analysis of the BELIV current transients in reverse and forward LIV biased junctions is presented. Simulation methods and illustrations of the simulated BELIV transients are also discussed. Principles of the junction profiling and of the photo-ionization spectroscopy by examination barrier capacitance transients are described to pave these applications of the BELIV technique. Afterwards, results of the specific BELIV applications are briefly illustrated such as: barrier evaluation for heterojunctions; fluence and temperature dependent BELIV characteristics in Si particle detectors; on-line barrier control during irradiation with hadrons; photoionization spectroscopy of deep traps in Si structures; profiling of junction location and thermal emission spectroscopy of industrial solar cells.

1. Measurement schemes and impact of the external circuit

1.1. Measurement circuitries

A generator of the linearly increasing voltage (LIV) is one of the main instruments for successful implementation of the technique. The output impedance of the LIV generator (GLIV) should be adjusted to a measurement circuit employed. The range of LIV generated pulse durations is desirable to be wide enough in order to implement different scan and spectroscopy regimes. The significant feature of the employed LIV generator would also be a possibility to combine and to vary the dc baseline voltage, - this is often necessary to govern dc biasing for a controllable shift of Fermi level in semiconductor material and for a priming of deep level filling. Also, a wide range of the output currents suitable for a reliable operation of a LIV generator is necessary for the measurements on the rather large area structures in order to make a reproducible charging of the large capacitance junctions. The essential characteristic of a real LIV generator is its linearity. As usual, the linearity can be verified by differentiating of a LIV pulse. This procedure highlights the initial curvature (due to the charge integration effects) of the rise to vertex component within a square-wave pulse. The change of linearity due to insufficient current supported by a LIV generator can also appear in further stages of a LIV pulse, if a resistance of a device under test unacceptably reduces with voltage increase.

The important role is inevitably played by the signal transfer lines, such as cables and connectors. The definite capacitance of coaxial cables determines a baseline square-wave signal. This signal should be evaluated and subtracted during trimming of the operating system in signal measurements. Also, the baseline capacitance appears in cryostats and sample mounting arrangements. This is an additional background capacitance, which also appears in a baseline square-wave signal and should be eliminated in precise measurements.

A sketch of measurement circuitry for implementation of different regimes, by employing a technique of barrier evaluation using LIV pulses (BELIV), is illustrated in Fig. 1.1. The measurement circuitry contains an adjusted output of a generator of linearly increasing voltage (LIV), a diode under investigation, and a load resistor connected in series. Current transients are registered using a 50 Ω external resistor or load input of the digital oscilloscopes. The 2 -4 channels of the digital oscilloscope are exploited for BELIV signal recording and for synchronous control of a linearity of a GLIV signal using a signal differentiating procedure installed within modern oscilloscopes, produced by Tektronix, Agilent and LeCroy. As mentioned, linearity of the GLIV signal is essential within implementations of the BELIV technique, therefore, several types of generators were tested to get suitable LIV characteristics.

The device under test (DUT) is commonly mounted on a cold finger within a vacuumed cryo-chamber in dark for examination of the temperature variations of BELIV characteristics, as sketched in Fig. 1.1a. A complementary continuous-wave infrared (IR) light is employed to vary expediently a filling of deep centres, and illumination is implemented by fiber transfer lines. The steady-state bias voltage can be additionally applied and varied to shift or stabilize Fermi level position.

Arrangement for the BELIV profiling is sketched in Fig.1.1b. The cross-sectional boundary of a layered junction structure is commonly scanned by a gold needle positioned by the precise 3-D stepper. The micro-structure of a needle tip determines an appearance of the spreading currents between the main plate electrode of a relatively large area and the needle tip on a boundary of a structure located (perpendicularly to the main plate electrode) within a definite layer and at definite depth point. The boundary of the samples for depth profiling should be carefully prepared to have a boundary surface as smooth as possible. Positioning of the probe is performed by a set of operations: raise a probe- its shift – relevant pressing of a probe at new position.

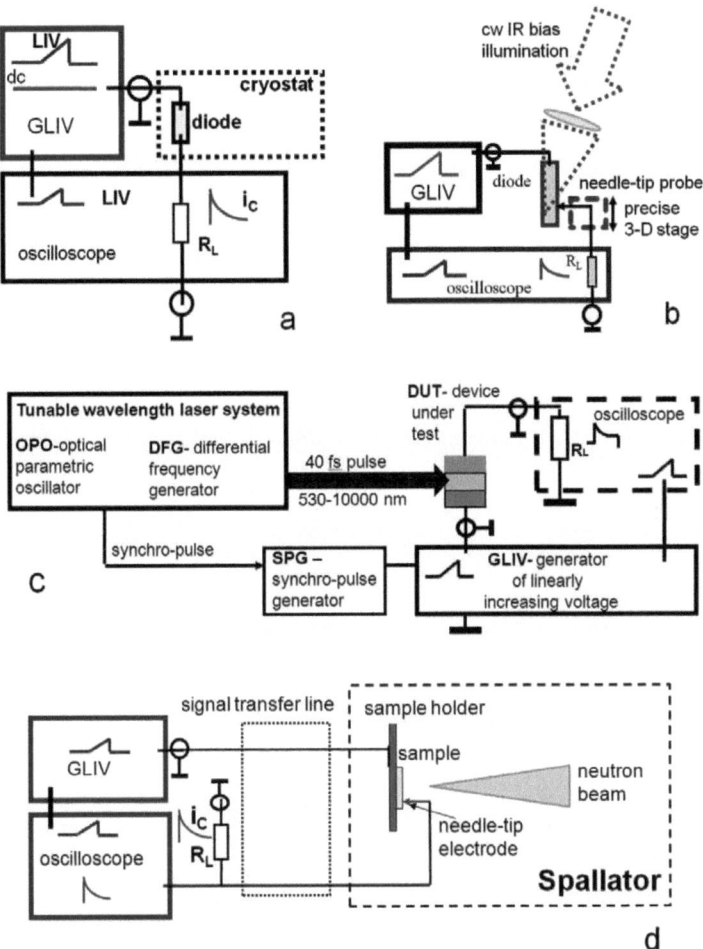

Figure 1.1. a- Sketch of the measurement circuitry for implementation of the temperature scans using BELIV technique. GLIV- generator of linearly increasing voltage. b- Measurement circuitry for profiling of doping and defects density. A continuous-wave (cw) infrared (IR) light source is employed to vary expediently a filling of deep centres. c- Measurement arrangement for deep level photo-ionization spectroscopy. d- Experimental arrangement for the on-line measurements of evolution of the radiation defects using BELIV technique.

A complementary continuous-wave light source, relevant for excitation of the electrically active traps, is employed to vary expediently a filling of deep centres.

The pulsed or steady-state illumination is implemented by laser or incandescent lamp light, focused onto examined probing spot or transferred by an optical fiber.

Spectroscopy of deep levels by combining the temperature scans with varied LIV pulse durations and analysis of the quantum energy dependent photo-ionization/ photo-neutralization responses is performed by using the measurement arrangements sketched in Fig. 1.1c. Photo-excitation in these spectral measurements is implemented by using the tuneable laser systems of parametric and differential frequency optical oscillators. As usual, short laser pulses in the range of 40 fs -30 ps of tuneable wavelength are employed.

The photocapacitance and photoconductivity responses and their relaxation lifetimes are controlled. These measurements are implemented at fixed or varied temperature. Recording of the photocapacitance changes within a rather long set of BELIV pulses before and after a single photo-excitation pulse, which is synchronized with a chain of the GLIV pulses, enables ones to unveil an evolution of the deep level filling and emptying processes, for a fixed excitation wavelength, and their characteristic lifetimes.

Evolution of radiation defects using BELIV technique is examined by placing of a mounted sample in dark within a beam of high energy particles. The measurement instrumentation is located in a safe area of the particle accelerator environment, as sketched in Fig. 1.1d. The longer transfer lines are necessary for such on-line experiments. Therefore, more careful primary evaluations of the baseline capacitance signals should be made and these signals must be eliminated in precise measurements. As usual, the BELIV transients are recorded every second during exposure within a particle beam. Examination of the changes of BELIV transient components enables ones to trace a reduction of the effective doping density (through barrier capacitance variations) and an increase of density of deep traps (via control of generation current component).

Samples for BELIV measurements are commonly mounted on a printed circuit board with a heat conductive substrate (for temperature scans) and with temperature sensors attached. The top needle-tip probes, supported by spring-pressure arrangements, are employed for structures with two-side electrodes.

It is worth noting that the central module (GLIV, DUT and oscilloscope) for BELIV measurements is rather simple. The central module can easily be combined with temperature stabilization and variation arrangements, with pulsed excitation and steady-state illumination light sources, with structure depth-scan probing instrumentation in order to perform the specially dedicated investigations. This module as alone instrument is suitable for the fast and simple estimations of quality of the barrier structures when measurements of rather small currents can be implemented. For the more comprehensive analysis, the influence of the external circuit should be evaluated.

1.2. Impact of the external circuit

In order to evaluate the role of the external circuit, which is rather important, behaviour of the circuit with inevitable elements (at least an RC chain) under action of the LIV pulses should be analyzed.

The solution of a simple differential equation $R_L di/dt+(1/C)i=A$, derived for the linear elements, as a capacitor C and an $R_L=R$ connected in serial, biased by a LIV pulse $U=At$, and using an initial condition of $i(t=0)=0$, leads to the expressions:

$$i(t) = AC(1 - e^{-\frac{t}{RC}}),$$

$$U_R(t) = ACR(1 - e^{-\frac{t}{RC}}) = \begin{cases} At|_{t \ll RC} \\ ACR|_{t \gg RC} \end{cases}, \qquad (1.1)$$

$$U_C(t) = A[t - RC(1 - e^{-\frac{t}{RC}})] = \begin{cases} A\dfrac{t^2}{2RC}|_{t \ll RC} \\ A(t - RC) \cong At|_{t \gg RC} \end{cases}.$$

Equations (1.1) imply that the linear voltage drop on a capacitor (and on a diode under test -DUT) appears for time instants $t \gg RC$.

The maximal barrier capacitance C_{b0} for DUT acts during the initial instants of LIV pulse (when $At < U_{bi}$). Thus, the fastest initial component of the BELIV current transient is determined by the transition time constant RC_{b0}. The linear RC_{b0} modifications (for instance, a shift) of the initial step of barrier capacitance charging current $i_C(t \approx 0)$ can be roughly emulated by a convolution integral

$$i_{RC}(t) = \frac{1}{\tau_{RC}} \int_0^t i_C(x) \exp[-\frac{(t-x)}{\tau_{RC}}]dx, \qquad (1.2)$$

shown by the broken curves in Fig. 1.2a. Then, the peak within a current transient appears on the $i_C(t)$ simulated curve (a dot curve in Fig. 1.2a).

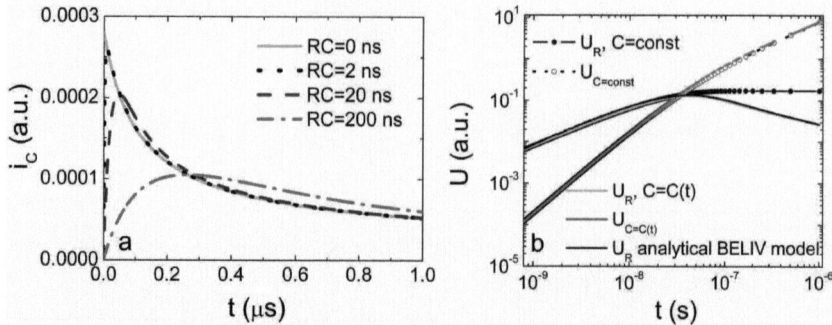

Figure 1.2. a- Simulated BELIV currents $i_C(t)$ for a diode with C_{b0} = 70 pF, calculated without delay (solid curve) and using convolution integral with RC values of RC= 2 ns (dot), RC= 20 ns (dash) and RC=200 ns (dash-dot). b- Simulated time dependent voltage drops on load resistor (R_L=50Ω) and on capacitor (C=430 pF) for LIV pulse of τ_{PL} =1 µs of a peak voltage U_P =8V. (After E. Gaubas et al, ISRN Materials Science, (2012) article ID543790, doi:10.5402/2012/543790, [12]).

A kink (at $t \approx RC$) within the initial rise front of the simulated BELIV transient and a peak position of the experimental one might be employed for the evaluation of the C_{b0} using a rough linear approximation (by a convolution integral) of a BELIV transient.

However, variation of the voltage drop $U_C(t)=A(t-RC(t))$ on a nonlinear DUT can be assumed being a linear function of t only in time scale $t>2RC$ (Eq. (1.1)). For comparison, the simulated (using Eqs. (1.2)) time dependent voltage drops on the load resistor (dot curve) and on a capacitor (solid curve), are illustrated in Fig. 1.2b. It is worth noting, that, for a dielectric capacitor within RC circuit, the voltage response, measured on the load resistor using LIV, is close to a square-wave pulse in the linear time scale.

The discussed estimation of the role of external circuit can be employed for a simple understanding of features of the registered BELIV transients. The interpretation of the BELIV transients can be simplified and the extraction of junction parameters can directly be performed if barrier charging currents prevail. However in real experimental situations, the more complicated interplay of processes within structure appears under LIV pulse of different polarity. Therefore, the detail analysis

14

and simulations of the BELIV transients are necessary for the more precise evaluation of barrier parameters.

2. Current transients in reverse and forward LIV biased junctions

A scale of durations of the LIV pulses employed in BELIV technique is ranged from 100 ns to 100 ms to get detectable and resolvable responses. This range is equivalent to the 10 Hz – 10 MHz frequency range in a wide-used TST measurement. For such a LIV pulse duration range, the fast components of the convection-displacement currents caused by movement of the field discontinuity at interface between depletion (DL) and electrically neutral (ENR) regions are really non-resolvable. The instantaneous shifts of DL-ENR interface are rather small under a linear voltage increase. Therefore, Ramo's currents of the durations of a few ns are rapidly relaxed. Then, any current value within a BELIV transient measured using reverse bias regime represents a current component ascribed to the instantaneous position of the DL-ENR interface, due to the drift-diffusion currents within a transitional layer. These current components can be assumed as equivalent to the steady-state process. The Ramo's current changes can only be observed as a noise signal overlapped with barrier charging current. This noise signal can appear if a LIV generator is not capable to supply a sufficient charge to electrodes. Thereby, the BELIV transients can be precisely described as a routine capacitance-voltage characteristic measured by TST, nevertheless, recorded rapidly, in accordance with a LIV pulse display. In the case of the diffusion (storage) capacitance analysis, the Ramo's current component can be a reason of the exponential current increase for the forward-biased junction.

2.1. Charge extraction BELIV regime

2.1.1. Trap free material

It has been shown [5], that analysis of convection-displacement currents can be implemented by consideration of the system capacitance changes in time, including capacitance dependence on field discontinuity location. Thereby, the basic relations for analysis of the BELIV transients are further derived employing barrier capacitance changes. As mentioned above, fast processes are stabilized for the considered range of LIV pulses. Therefore, the main features of the BELIV transients can be revealed within depletion approximation.

The BELIV technique for a reverse biased diode is based on analysis of the changes of barrier capacitance (C_b) with linearly increasing voltage $U_p(t)=At$ pulse. The $C_b(t)$ dependence on voltage $U_p(t)$ and thereby on time t can be described using the depletion approximation [7], for charge extraction in the trap-free material. This approximation for an abrupt p$^+$-n junction in diode leads to a simple relation $C_b = C_{b0}(1+U/U_{bi})^{-1/2}$, where barrier capacitance for a non-biased diode of an area S is $C_{b0} =\varepsilon\varepsilon_0 S/w_0=(\varepsilon\varepsilon_0 S^2 eN_D /2U_{bi})^{1/2}$. The other symbols represent: ε_0 is a vacuum permittivity, ε material dielectric permittivity, e elementary charge, U_{bi} built-in potential barrier, $w_0 = (2\varepsilon\varepsilon_0 U_{bi}/eN_D)^{1/2}$ width of depletion for the non-biased junction, $A=U_p/\tau_{PL}$ ramp of LIV pulse with U_P peak amplitude and τ_{PL} duration.

The depletion approximation [7] seems to be a rather good model for description of the charge extraction transients in time scale t longer than the dielectric relaxation $\tau_M=\varepsilon\varepsilon_0/e\mu n_0$ time, when employing LIV pulses of durations $\tau_{PL}>\tau_M$. Here, it is assumed that an equilibrium carrier density n_0 is equal to the effective doping density ($n_0=N_{Deff}$). This is maintained by a short Debye length $L_D=(2\varepsilon\varepsilon_0 kT/e^2 n_0)^{1/2}$ relatively to the characteristic depletion widths (w_0, $w=w_0(1+U/U_{bi})^{1/2}$) and a geometric thickness d ($L_D \ll w_0$, w, d) for $N_{Deff} > 10^{11}$ cm^{-3}. Precision can be increased by taking into account a free carrier tail within a transitional layer [7]. Then, a barrier capacitance is corrected by a voltage term:

$$U* = \frac{[U - \frac{kT}{e}(1 - \exp(-\frac{eU}{kT}))]}{[(1 - \exp(-\frac{eU}{kT}))]^2} \quad , \qquad (2.1)$$

relative to the depletion approximation, where the total voltage across the depletion region U should be replaced by U^* for the exact solution. For practical applications in the range of $U>U_{bi}$ and $kT/e \ll U$, the exponential terms can be neglected, and correction becomes of the order $U^*=U-kT/e$.

The time dependent changes of charge $q=C_b U$ within junction determine a current transient $i_C(t)$ in the external circuit, expressed using depletion approximation which gives solutions with precision of thermal potential, as:

$$i_C(t) = \frac{dq}{dt} = \begin{cases} \dfrac{\partial U}{\partial t}\left(C_b + U\dfrac{\partial C_b}{\partial U}\right) = \dfrac{\partial U_C}{\partial t}C_{b0}\dfrac{1+\dfrac{U_C(t)}{2U_{bi}}}{\left(1+\dfrac{U_C(t)}{U_{bi}}\right)^{3/2}} \approx AC_{b0}\dfrac{1+\dfrac{At}{2U_{bi}}}{\left(1+\dfrac{At}{U_{bi}}\right)^{3/2}}, \quad at \quad U < U_{FD} \\[3em] \dfrac{\partial U}{\partial t}\left(C_{geom} + U\dfrac{\partial C_{geom}}{\partial U}\right) = \dfrac{\partial U_C}{\partial t}C_{geom} \approx AC_{geom}, \quad for \quad U \geq U_{FD} \end{cases}$$

$$(2.2)$$

This transient for a rather small peak voltage U_P below values $U_P < U_{FD}$ of full depletion voltage (U_{FD}) contains an initial ($t=0$) step AC_{b0} due to displacement current and a descending component governed by the charge extraction. For an insulating material and for $U_P > U_{FD}$, this transient contains only a displacement current step. When parameters of material and of LIV pulse match a condition $0 < U_P \leq U_{FD}$, the transient contains a displacement current step and a descending charge extraction component, which saturates at time instant $t_{FD} = U_{FD}/A + C_{geom}R_L$.

The descending component of charge extraction for $U_P < U_{FD}$, gives an additional relation to extract U_{bi} by taking a ratio $r_m = i_C(0)/i_C(t_m) \geq 1$ at fixed time instant t_m. Actually, a real positive root of the cubic equation $(A/U_{bi})^3 t_m^3 + (3 - r_m^2/4)(A/U_{bi})^2 t_m^2 + (A/U_{bi})[3 - r_m^2]t_m + (1 - r_m^2) = 0$ should be found. The Viete's relation for a product of roots of the cubic equation and Routh-Hurwitz' criteria (for $(1 - r_m^2) < 0$) indicate that such a root exists. Subsequently value of N_D is evaluated by substituting the extracted U_{bi} within the initial current expression $i_C(0) = AC_{b0}$. Having determined N_D, the density of acceptors N_A in p^+ layer can be verified by using a well-known relation $U_{bi} = (kT/e)\ln(N_A N_D/n_i^2)$ with intrinsic carrier density n_i. Therefore, the mentioned procedure, using Eq. (2.2), can be employed for primary analysis of a fragment of transient (in perfect junction structures), experimentally measured for a fixed LIV pulse duration τ_{PL}. The experimental BELIV transients, measured on a non-irradiated diode and shown in Fig. 2.1, are in agreement with diode current transients approximated by equation Eq. (2.2). The vertex amplitude (Fig. 2.1) within a transient is proportional to a value of a ramp A of LIV pulse, which is controlled by differentiating $(dU_p/dt) = A$ the LIV pulse ($U_p(t)$), as illustrated in Fig. 2.1. The simulated (using Eqs. (2.2 and 1.2)) BELIV current transient is illustrated by solid curve in Fig. 1.2(b). To extract the barrier parameters more precisely, the above mentioned procedures can be applied after diffusion current is evaluated. The latter may be a reason in formation of a pedestal for $i_C(t)$ changes.

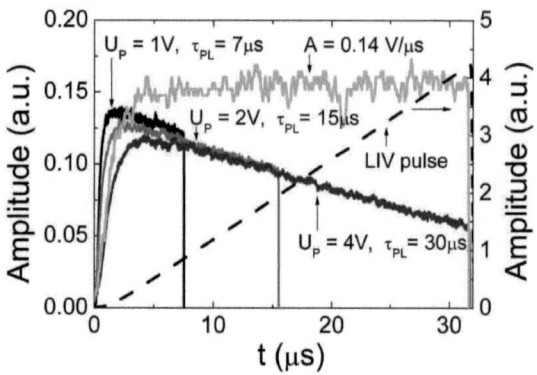

Figure 2.1. Transients of the charge extraction current measured in the commercial diode at a reverse (U_R) polarity of LIV pulses of varied duration keeping a constant LIV ramp. A dashed curve represents an experimental LIV pulse while the grey curve is a derivative in time of this LIV pulse. (After E. Gaubas et al, ISRN Materials Science, (2012) article ID543790, doi:10.5402/2012/543790, [12]).

The diffusion current $i_{diff}(t) = i_{diff\infty}[1-e^{-eAt/kT}] \cong eSn_i^2 L_{Dp}/N_D\tau_p[1-e^{-eAt/kT}] \approx eSn_i^2 L_{Dp}/N_D\tau_p$ is a rapidly stabilized (for $t > k_BT/eA$) function at reverse biasing. Here, additional symbols represent: kT is thermal energy at temperature T; $L_{n,p} = (D_{n,p}\tau_{n,p})^{1/2}$ is a diffusion length for electrons (n) and holes (p) in p and n layers of a diode, respectively. The stabilized value of $i_{diff}(t >> kT/eA) = i_{diff\infty}$ can be ignored in comparison with $i_C(t)$ for properly fabricated (containing low density of traps and large $L_{n,p}$) diodes in the realistic range of LIV pulse durations. The diffusion current in the range of $t \lesssim kT/eA$ determines a differential resistance of a junction, which may limit an increase of $i_C(t \approx 0)$.

Really, a delay appears due to serial processes of dielectric relaxation within quasi-neutral range of non-depleted n-layer, drift and diffusion of carriers to complete a circuit. The initial delay is caused by the characteristic times of diffusion $\tau_D = L_D^2/2D$ and of dielectric relaxation τ_M. The mentioned characteristic times, summarized as τ_{RC}, may comprise a delay. Additional RC appears due to the external circuit and the initial non-linearity of GLIV pulse. The latter can be noticed as a deviation from a square-wave shape pulse within very initial stages of the differentiated experimental GLIV pulse.

The BELIV transients for the abrupt junction diodes have been discussed above. Similarly, for a linearly graded (L_g) junction, the current $i_C(t)$, due to depletion of L_g junction, can be expressed as $i_{CLg}(t)=AC_{b0Lg}(1+2At/3U_{biLg})/(1+At/U_{biLg})^{4/3}$. Here, the re-arranged parameters of U_{biLg}, w_{0Lg} and C_{b0Lg}, as e.g. in [2, 10], should be used.

As shown above, the BELIV technique is suitable for the direct extraction of barrier parameters of the properly fabricated junctions using the reverse bias regime. In order to increase precision of the parameter extraction procedure when using a rather wide range of applied voltages and of junction areas, the modifications of BELIV transients should be simulated including a voltage sharing within a measurement circuit.

2.1.2 Simulations of the impact of the external circuit

The load resistor R_L, as a linear circuit component for registration of the BELIV current transient, determines the time-dependent voltage sharing in RC circuit between the R_L and the diode with barrier capacitance C_{b0}.

However, variation of the voltage drop $U_C(t)=A(t-RC(t))$ on a nonlinear device under test (DUT) can be assumed being a linear function of t only in time scale $t>2RC$ (Eq. (1.2)). For comparison, the simulated (using Eq. (1.2)) time dependent voltage drops on the load resistor (dot curve) and on a capacitor (solid curve), are illustrated in Fig. 1.2b.

For the more precise description of BELIV transient, a generalized differential equation

$$\frac{dU_C(t)}{dt} \frac{(1+\frac{U_C(t)}{2U_{bi}})}{(1+\frac{U_C(t)}{U_{bi}})^{3/2}} - \frac{U_p(t)-U_C(t)}{RC_{b0}} = 0 \qquad (2.3)$$

with the initial conditions $U_p(t=0)=0$ and $U_c(t=0)=0$ should be solved to determine $U_C(t)$. This equation (Eq. 2.3) is derived assuming the time-dependent voltage drops on DUT, as $U_C(t)=U_p(t)-U_R(t)$.

This Eq. (2.3) is a non-linear differential equation, more complicated than a Riccati' one. Only the numerical solutions of the Eq. (2.3) can be obtained. Using a solution $U_C^*(t)$ of equation Eq. (2.3), a voltage drop on the load resistor, as $U_R^*(t)$ $=U_p(t)- U_C^*(t)$, represents the simulated BELIV transient. Thus, a fitting procedure, of the numerically simulated BELIV transients $U_R^*(t)$ to the experimental ones, is inevitable and has been employed in this work for precise extraction of the junction parameters.

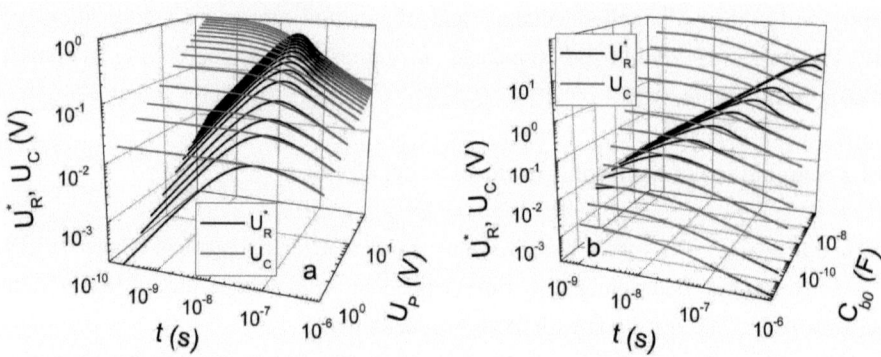

Figure 2.2. a- Numerically simulated BELIV voltage transients $U_R^*(t)$ as a function of a LIV pulse peak voltage U_P compared with those simulated by analytical approximation. b- Numerically simulated BELIV voltage transients $U_R^*(t)$ as a function of initial barrier capacitance values C_{b0} compared with those simulated by analytical approximation. (After E.Gaubas et al, ISRN Materials Science, (2012) article ID543790, doi:10.5402/2012/543790, [12]).

Comparison of the simulated voltage transients (using $U_R^*(t)$ obtained by solution of Eq. (2.3), and using $U_C(t)$ (Eq. (2.2)) values evaluated for fixed C using analytical approximations (Eqs. (1.1) and (1.2) for $RC=20$ ns) are presented as a function of LIV pulse peak voltage U_P and as a function of value of the equilibrium barrier capacitance C_{b0} in Figs. 2.2 (a and b), respectively. It can be noticed in Figs. 2.2(a, b) that a deviation from the analytically simulated curves appears in the range of the BELIV voltage peak. This deviation increases with increment of U_P and of C_{b0} values. These deviations can be explained by the relative enhancement of barrier charging current through the load resistor. This current increase modifies non-linearly the voltage drops (as included in Eq. (2.3)) on a load resistor and on a diode. Thus, the analytical approximation (Eq. (1.2)) can be exploited for primary analysis of the BELIV transients only in the range of small i_C currents, and including RC shift.

It can be deduced from Fig. 2.2 that a noticeable impact of voltage sharing appears for reverse voltages above 10 V and values of DUT capacitance values above 1 nF for barrier structures with $U_{bi}<1$ V, where the shape of a BELIV transient is modified. However, presence of traps within junction layers also can significantly change a shape of BELIV transients.

2.1.3. BELIV transients in trap containing material

Variety of processes and effects, determined by different ratios of the characteristic times of carrier capture as well as of carrier thermal emission ascribed to deep traps, and of dielectric relaxation, responsible for a stabilized depletion boundary, is rather wide [6], especially when carrier redistribution among several centres takes place. Here, an impact of only a few of them is briefly discussed.

2.1.3.1. Generation centres

The deep traps are responsible for generation current within depletion region. Depending on carrier emission (τ_{em}) and capture (τ_{cp}) characteristic times (when $\tau_{em}>\tau_{cp}>\tau_{PL}$) a definite density of traps can persist filling for a rather long time-gap relatively to a LIV pulse duration.

Generation centres can be observable in BELIV current response either by modification of a depletion width (by changing of the applied electric field distribution) during a LIV pulse or by collected charge, when increment of depletion width (bulk) highlights an impact of the generation current. The prevailing regime can be resolved only in the experiments.

Modification of the depletion width (when generation centres are able to redistribute electric field within a depletion area) can appear through changes in the built-in potential as $U_{bi}=(kT/e)\{\ln(N_A N_D/n_i^2)+\ln[1-(N_{An}\pm(N_T^{\pm}-n_T(t)))/N_D]\}$, where N_D is a doping density of the non-irradiated material and N_{An} is a density of radiation induced acceptors in the n-type material within a band-gap. Also, there are traps $N_T^{\pm}$ of acceptor (+) or donor–type (-) within the upper-half of the band-gap with respective temporal their filling $n_T(t)$ (here, superscript (+/-) shows selection of the sign before bracket, not electrical charge). Thus, the built-in barrier has two components $U_{bi}(t) = U_{bi0} + (kT/e)\{ln[1-(N_{An}\pm(N_T^{\pm} - n_T(t)))/N_D]\}$. The second component can be modulated by LIV, when the duration of the characteristic emission/capture processes approaches to a dielectric relaxation (τ_M) time, i.e. $\tau_{cp} \approx \tau_{em} \sim \tau_M$. Also a component in the temporal modulation of the depletion width $w(t)$ appears due to changes in the effective doping as $N_{Deff}(t) = N_D \times [1-(N_{An}\pm(N_T^{\pm} - n_T(t)))/N_D]$. These fast trap filling variations result in temporal changes of barrier capacitance as

$$C(t) = C_{b0} \times \left(1 - \frac{N_{An} \pm (N_T^{\pm} - n_T(t))}{N_D}\right)^{1/2} \times$$

$$\times \left[1 + \frac{U(t)}{U_{bi0}}\left(1 + \frac{kT}{eU(t)}\ln(1 - (N_{An} \pm (N_T^{\pm} - n_T(t))/N_D))\right)\right]^{-1/2} \cdot \tag{2.4}$$

Then, the BELIV current can be derived as a time differentiated response:

$$i(t) = \frac{dU_C}{dt} \times \left[C(t) + U_C(t)\left[\frac{dC}{dw}\frac{dw}{dU_C} + \frac{dC}{dw}\frac{dw}{dn_T}\frac{dn_T}{dU_C}\right]\right]. \tag{2.5}$$

However, this leads to very cumbersome expressions. The simulated BELIV current transient, using the approach of fast traps ($\tau_{cp} \approx \tau_{em} \sim \tau_M$), is compared in Fig. 2.3 with that for a diode without traps, when keeping the same values of other parameters. The simulated transient shows an impact of carrier generation current in the ulterior stages of a transient by exceeding those current values for a diode without traps. The main point of reference for separation of this trap modulated BELIV regime is a coincidence of current values within initial stage of transients for a fixed LIV pulse ramp dU_C/dt. Although, the simulated transients appear to be considerably complicated when variation of the parameters of N_{An}, $N_T^{\pm}$, $n_T(t)$ is involved. It is worth noting that this regime should be more pronounced when a single type (fast) trap dominates.

For charge collection BELIV regime, the generation current is included by a simple increase of volume from which carriers are collected by increased depletion width, during LIV pulse evolution. The generation current

$i_g(t) = en_i Sw_0 (1+U_C(t)/U_{bi})^{1/2}/\tau_g$ increases with voltage $U_C(t)$ and can exceed the barrier charging current in the rearward phase of the transient. Here, N_D in expressions for w_0 and C_{b0} should be replaced by its effective value $N_{Def} = N_D \pm N_T^{\pm}$, due to (compensation) charged traps of density N_T. Then, transient of the total reverse current is described by a sum of the currents:

$$i_{R\Sigma}(t) = i_C(t) + i_{diff}(t) + i_g(t) =$$

$$= AC_{b0} \frac{1 + \dfrac{U_C(t)}{2U_{bi}}}{(1 + \dfrac{U_C(t)}{U_{bi}})^{3/2}} + i_{diff\infty}(1 - e^{-\frac{eU_C(t)}{kT}}) + \frac{en_i Sw_0}{\tau_g}(1 + \frac{U_C(t)}{U_{bi}})^{1/2} \cdot \quad (2.6)$$

The descending component of the charge extraction and the ascending component of the generation current imply existence of a current minimum within a current transient (approximated by Eq. (2.6)). The time instant t_e, for which this extremum appears, is determined by using a condition $di_{R\Sigma}/dt|_{t_e} = 0$. This leads (at assumption $di_{diff}(t)/dt = 0$ for $t_e >> kT/e(dU_c(t)/dt)$ to a relation for the peculiar time instant t_e (for $i_{\Sigma}(t) = \min$), as

$$t_e = \frac{U_{bi}}{\dfrac{dU_C}{dt}i_g(0)}[\frac{i_C(t)}{4} - i_g(0) + \sqrt{(\frac{i_c(t)}{4})^2 + \frac{3}{2}i_c(t))i_g(0)}]. \quad (2.7)$$

The initial component of the composite current $(i_{R\Sigma}(t) \approx i_C(t) + i_{diff}(t) >> i_g(0)$ for $t << t_e$) can be exploited for evaluation of the barrier U_{bi} height. Subsequently, carrier generation lifetime can be evaluated by using Eq. (2.7). The simulated current transient using Eq. (2.6) for the latter BELIV (charge collection) regime is illustrated in Fig. 2.3. Also, the peculiar points are denoted in this Fig. 2.3. The main point of reference, to identify this charge collection BELIV regime, is an increment of current value relatively to that for the non-irradiated sample at fixed LIV pulse ramp dU_c/dt. This regime is the most probable one when several traps (primary filled) of different species simultaneously emit carriers.

For instance, in analysis of the heavily irradiated diodes, nearly all the "native" carriers ($n_0 = N_D$) are captured by N_T traps. Capture of n_0 carriers leads to a full depletion regime. In the fully depleted diode, enhancement of applied voltage determines a shortening of the time of carrier transit $\tau_{tr} = d^2/\mu U_C|_{FD}$ across the inter-electrode gap equal to sample thickness d, approaching to relation $\tau_{tr} \leq \tau_{em}$. In diode biased above full depletion voltages, the total BELIV current, can be described (similarly to the method used in [7]) by considering conductivity and displacement current components. This consideration is performed for variable voltage and for surface charge changes on electrodes, at standard boundary conditions (for electric field $E|_{x=d} = 0$, $dE/dx|_{x=d} = 0$ and for potential $V|_{x=d} = 0$). This solution is expressed and approximated (for $\tau_{tr} \sim \tau_{em}$ and $n \approx n_0 = N_D$), as

$$i(t)\big|_{FD} = eS\frac{d}{2}(\frac{dn}{dt}+\frac{dp}{dt})+\frac{\varepsilon\varepsilon_0 S}{d}\frac{dU_C}{dt} \approx$$
$$eS\frac{d}{2}\frac{n_0}{\tau_{tr}}+C_{geom}\frac{dU_C}{dt} = e\frac{S}{2d}n_0\mu_n U_C(t)+C_{geom}\frac{dU_C(t)}{dt} \cdot \qquad (2.8)$$

Here, $C_{geom} = \varepsilon_0\varepsilon S/d$ is a geometrical capacitance. It is clear that for trap free insulating material (when $d(n,p)/dt=0$) the BELIV current transient $i_{CFD}=C_{geom}(dU_C/dt)$ acquires a shape of square-wave pulse, at voltages $U_C|_{FD}$.

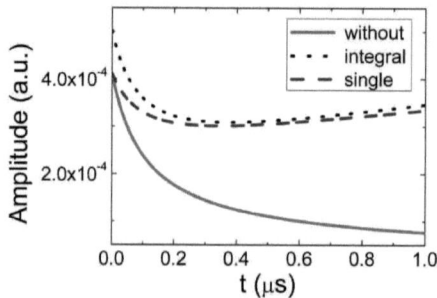

Figure 2.3. Comparison of charge extraction BELIV current transients simulated for single type traps (dash) and for simultaneously acting several type generation centres (dots) with that simulated without contribution of thermal emission from traps (solid curve). (After E.Gaubas et al, ISRN Materials Science, (2012) article ID543790, doi:10.5402/2012/543790, [12]).

For the traps rich and compensated material, further increment of voltage above U_{FD} leads to the increase of current component $i_{emFD}(t)=e(S/2d)n_0\mu_n U_C(t)$, added to a i_{CFD}.

To resolve the prevailing BELIV regime, the external factors (e.g. steady-state light biasing, dc pedestal, temperature, etc.) can be varied. These factors modify the occupation of the trap states and highlight the dominant components of current.

2.1.3.2. Carrier capture centres

In a reverse biased diode, the carrier capture process is probable and makes the main impact either within a Debye screening length (for a steady state) or during a dielectric relaxation time τ_M within a transition distance $\lambda = [2\varepsilon_0\varepsilon(E_F-E_T(\lambda)/e^2 N_D]^{1/2}$ at a depletion boundary [7]. The transition distance λ defines a length at which flat band condition (going from the depletion region, i.e. band bending side) for a definite centre E_{TC} is achieved. This is reached by a voltage drop $\Delta U_\lambda=(E_F-E_T(\lambda))/e$ on a λ thick depletion layer close to the interface with electrically neutral material region. This voltage drop ΔU_λ positions the trap level below a Fermi level (trap is occupied and neutral). The mentioned definition of λ can be re-arranged to a condition within a scale of characteristic times, as $\tau_M=\tau_{tr\lambda}/2$, with $\tau_M=\varepsilon\varepsilon_0/e\mu n_0$ and $\tau_{tr\lambda}=\lambda^2/\mu\Delta U_\lambda$. The latter condition tells that λ is a distance for which dielectric relaxation time should be a half of the transition time ($\tau_{tr\lambda}/2$). This result can be understood by the interplay of conductivity and displacement current components within a total current (like the first component ($d/2$) of a sum in Eq. (2.8), when only a half of the conductivity current flows towards external circuit). On the other hand, screening of the depletion field within λ can be reached by extraction of thermally emitted carriers from the depleted layer and by free carriers diffused from the side of electrically neutral material,- this shortens a relaxation time twice. Recombination or capture of carriers within λ layer modifies this screening condition as $\tau_M=[2/\tau_{tr\lambda}+1/\tau_R]^{-1}$. Thus, transit time is modified due to carrier capture within λ layer, as the carrier diffusion length (from a neutral material side) is reduced (relatively to that of $\tau_R \to\infty$). Then dielectric relaxation time (during which an impact of free carriers on space charge and electric field vanishes at a depletion boundary) approaches to the carrier capture time ($\tau_M\to\tau_{cp,R}$), when the transit time becomes significantly longer than that of capture/recombination ($\tau_{cp,R}$) one. In fully depleted (insulating) material, the carrier capture and recombination processes are essential nearby the electrodes, although processes are of the same origin.

In material containing high density of wide spectrum of deep traps N_T, current $i_{cp}(t)$ generated by GLIV flows to fill the traps. Then, position of Fermi quasi-level varies within band-bending λ layer of the depletion range. The current response (due to carrier capture within λ layer in time-scale of a LIV pulse) can be expressed as

$$i_{cp}(t) = \frac{en_0(t, \tau_M, \tau_{tr\lambda})S\lambda}{\tau_{cp}},$$

(2.9)

when several species of deep centres are involved. Alternatively, carrier recombination/capture processes can be considered through capture rate $u(t)$ (accepted in Ref. [7] as the so called extended depletion approximation). Carrier capture actually modifies the doping $N_D(t)$ density capable to ensure the barrier, $N_D(t)=N_{De}u(t)$. For high density of carrier capture centres, an amplitude of response is nearly proportional to the current $i_{cp}(t)$. This carrier capture ($u<1$) response saturates and approaches to value $u(t>\tau_{cp})=1$ for time instants exceeding carrier capture time, i.e. when all the carrier capture centres are filled. For relatively long LIV pulses, with $\tau_{PL}>>\tau_{cp}$ and with small value of ramp A, an external current is insufficient to fill traps during the initial stages of the LIV pulse. The simulated BELIV current transients, using approximations involved within derivation of Eqs. (2.6 and 2.8), are illustrated in Fig. 2.4.

The simulated BELIV current transients indicate that clear barrier capacitance (initial peak due to barrier capacitance charging current) can be observed when trap density is lower than a density of dopants. Only ascending current component can be observed for trap density higher than that of dopants.

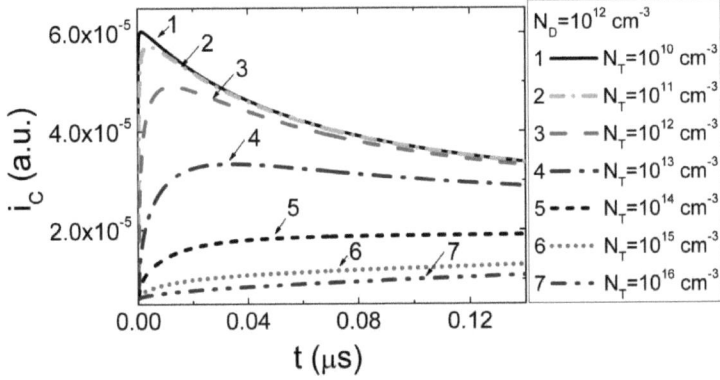

Figure 2.4. Comparison of the simulated charge extraction BELIV current transients varying density (10^{10} -10^{16} cm^{-3}) of carrier capture centres in Si material with N_D=10^{12} cm^{-3}. (After E. Gaubas et al, ISRN Materials Science, (2012) article ID543790, doi:10.5402/2012/543790, [12]).

It has been illustrated that traps considerably complicate the interpretation of the changes of BELIV transient shape. In order to identify these traps, the comprehensive investigations of temperature and of bias illumination spectrum dependent variations of BELIV transients should be combined. The BELIV regimes and techniques for implementation of such measurements are illustrated in Sections from 5 to 10. The combining of the reverse and forward biasing regimes of BELIV technique can also be fruitful for characterization of barrier.

2.2. Charge injection BELIV regime

Transients of the barrier charging current and their analysis become more complicated at forward biasing of a diode (a regime of minority carrier injection). Then, a voltage drop on a load resistor should be always included: $U_F(t)=At-R_L i_{\Sigma F}(t)$, to avoid the phantom roots and poles within model equations. Generally, current appears to be composed of the barrier ($i_{CF}(t)$) and of the diffusion (storage) capacitance charging ($i_{Cdiff}(t)$) currents and of recombination ($i_{RC}(t)$) and injection/diffusion ($i_{injcF}(t)$, $i_{difF}(t)$) ones, as

$$i_{\Sigma F}(t) = i_{CF}(t) + i_{Cdiff}(t) + i_R(t) + i_{injcF}(t) =$$

$$= \frac{\partial U_F(t)}{\partial t}\left[C_{b0}\frac{(1-\frac{U_F(t)}{2U_{bi}})}{(1-\frac{U_F(t)}{U_{bi}})^{3/2}} + C_{diff0}(e^{\frac{eU_F(t)}{k_BT}}(\frac{eU}{k_BT}+1)-1)\right]+$$

$$+\frac{en_iw_0S}{2\tau_R}(1-\frac{U_F(t)}{U_{bi}})^{1/2}e^{eU_F(t)/2k_BT)} + i_{diff0}(e^{eU_F(t)/k_BT)}-1) \quad .$$

$$(2.10)$$

Here, additional symbols represent: $\tau_R=[1+2(n_i/n_0)\cosh(E_T-E_i)]/\sigma_c v_T N_T \approx 1/\sigma_c v_T N_T$, $i_{diff0}=eSn_i^2 D_p^{1/2}/N_D \tau_p^{1/2}$, $C_{diff0}=i_{diff0}/(dU/dt)\cong i_{diff0}/A$. The storage (diffusion) capacitance is routinely assumed to be $C_{diff}=d(\int_0^t i_{diff0}(\exp(eU(t')/kT)-1)dt')/dU$. Actually, it can be derived on the basis of Ramo's theorem. In can be deduced from [5] that exponential components in the integral for storage capacitance are similar to those derived by including image charges supplied by an external voltage source. However, the process of the storage capacitance charging appears to be significantly slower, as it can be estimated from a comparison of Ramo's current and the diffusion pulse durations [5, 11]. At forward bias LIV, the initial stage of the BELIV transient is governed by a barrier capacitance current (first constituent in equation Eq. (2.10)), which rather slowly increases with $U_F(t)$. At $t\cong0$, i.e. for $eU_F(t)/kT \approx 0$, diffusion capacitance and diffusion current are close to zero. Thus, at low injection level regime, the barrier capacitance component prevails over the diffusion and recombination current constituents within initial stage of the current transient. The geometrical capacitance determines the initial step and capacitor like BELIV transient behaviour, when a diode is fully depleted without external voltage. However, contrary to reverse biasing (Eqs. (2.2- 2.6), where charge extraction causes a negative constituent U_RdC_b/dU_R), the barrier capacitance current component shows a positive derivative $U_FdC_b/dU_F>0$ for the forward biased diode. This is determined by a denominator function within the first constituent of Eq. (2.10). The amplitude of the BELIV current, during rearward phase of BELIV current transient, exponentially increases due to i_{Cdif} and i_{dif} for the short LIV pulses, (Fig. 2.5a). Thereby both the barrier capacitance ($i_{CF}(t)$) charging current and the total current ($i_{\Sigma F}(t)$) increase with LIV. Actual voltage drop on a diode $U_F(t)=At -R_L i_{\Sigma F}(t)$ deviates from a linear increase, when $i_{\Sigma F}(t)$ is enhanced, even in the case when GLIV signal is perfect. This requires a solution of voltage sharing equation similar to that of Eq. (2.5). Then, the initial increase of $i_{CF}(t)$ is additionally modified by the RC_{b0} parameter of a circuit. Actually, only numerical simulations of time-dependent current $i_F(t)$ variations should be employed for the precise evaluations of barrier parameters at elevated voltages.

The simulated variations of the current components and of the total current (during LIV pulse) are presented in Fig. 2.5a. These transients have been simulated by assuming a long carrier lifetime approximation. The two componential $i_{\Sigma F}(t)= i_{CF}(t)+ i_{Cdif.}(t)$ current transients are clearly observed in Fig. 2.5a. The storage capacitance component dominates within $i_{\Sigma F}(t)$ during ulterior stages of a transient. Small amplitudes of $U_F < U_{bi}$ are preferential in simulations and experiments, in order to reduce a non-linear sharing of voltage drops between the load resistor and the device under test.

The transients of the composite (total) current are illustrated in Fig. 2.5b, those were simulated for two values of U_F by keeping the same pulse duration (i.e. varying ramp) and including the initial delay of $RC=20$ ns. It can be noticed that the initial amplitude and, especially, the rearward component of a transient is rather sensitive to the LIV pulse ramp. The experimental BELIV current transients obtained by varying the ramp of forward bias are illustrated in Fig. 2.5c. The qualitative comparison of the simulated and experimental changes of the amplitude and of the shape within BELIV transients have been made using a set of BELIV transients measured on Si diode irradiated with rather small fluence (Fig. 2.5c). The qualitative changes of the shape of these experimental BELIV transients as a function of LIV voltage (LIV pulse duration) are in good agreement with those simulated (Fig. 2.5b) using approximation of long carrier lifetimes.

For quantitative estimation of barrier characteristics, the combined analysis of height of the initial step of current for the reverse $i_{R\Sigma}(0) \approx AC_{b0} + i_g$ and for the forward $i_{\Sigma F}(0)= AC_{b0} + (eSn_i w_0/2\tau_R)$ biased diode can be performed.

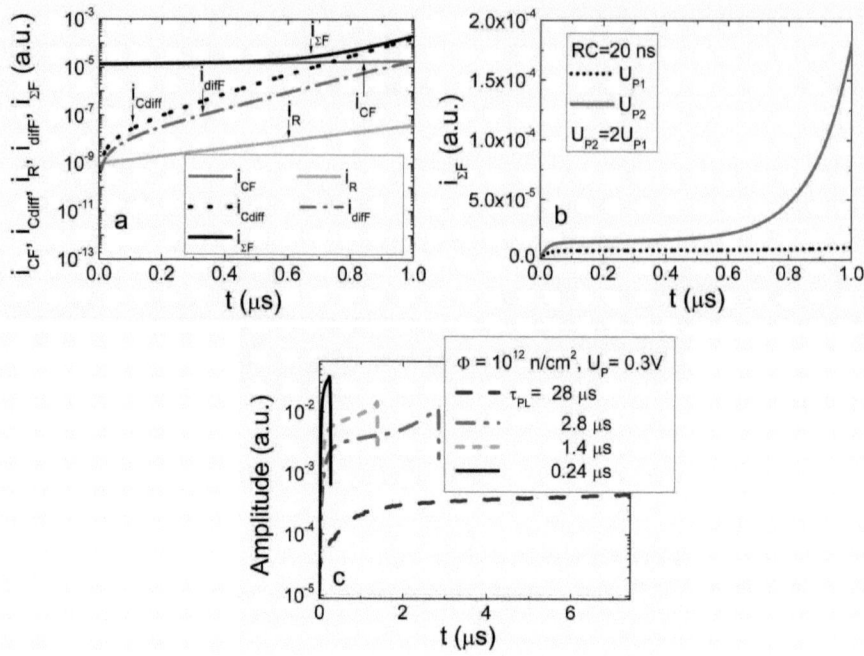

Figure 2.5. a- Simulated total current (black solid) at forward LIV bias composed of barrier (grey solid) and of storage capacitance (dot) currents as well as of recombination (light grey solid) and of diffusion (dash-dot) currents. b - The simulated BELIV transients varying U_P of the forward LIV bias. c- Experimental BELIV transients for forward (U_F) biased diode varying ramp A of LIV pulses. The pulse peak amplitude U_P =0.3 V was kept invariable while pulse duration was varied. (After E. Gaubas et al, ISRN Materials Science, (2012) article ID543790, doi:10.5402/2012/543790, [12]).

Combining of the reverse and forward bias BELIV regimes is expedient in profiling of the multilayered junction structures. The change of conductivity type and non-linearity of the probe electrodes can then be evaluated more reliably.

3. Principles of the profiling of layered junction structures

BELIV technique can be suitable for scanning of interfaces and of definite layer characteristics in multi-layered structures [12, 13]. It is based on analysis of the shape of BELIV transients inherent for different layers. Electrodes and interfaces between layers, comprising the ohmic conductivity, show a transient which repeats a shape of a LIV pulse. While, junctions exhibits barrier capacitance inherent current transients those also represent biasing polarity. Geometrical characteristics, such as electrode area and its geometrical extent can also be unveiled from the analysis of BELIV transient shape.

The characteristics, discussed in Section 2, are inherent for the measured BELIV transients on structures with parallel-plate layers and electrodes. For electrodes of the perpendicular configuration, using a needle-tip electrode positioned on the cross-sectional boundary of a parallel-plate layered structure, the current spreading effect should be included. The spreading of current trajectories acts as a serial resistor R_S introduced into BELIV measurement circuitry. For perpendicular geometry of electrodes, the impact of the interface at the needle-electrode is important. This interface can be either ohmic or injecting.

The layered junction structures can be controlled by profiling of barrier charging currents for the perpendicularly located electrodes. The boundary needle-tip electrode introduces several complications when using a simple evaluation of parameters. The spreading of currents from a nearly point contact appears even in the case when a single homogeneously doped layer is profiled. The simulated (using TCAD platform) distribution of potential and of current density are illustrated in Figs. 3.1a and 3.1b, respectively. Actually, the needle electrode induces a contact of small surface area for a current flow. This is a reason for a spreading resistance ($R_{S\perp}$) dependent on probe location y and on electrode diameter a, which can be expressed by the geometrical (y, a, Θ) and voltage (U) parameters, as:

$$R_{S\perp} = \rho(U, y+a)\frac{l(y,\Theta)}{S(y,\Theta)} = \rho_0(y+a)\frac{l_{eff}}{S_{eff}}(y,t,U(t)) \quad . \quad (3.1)$$

Here, ρ is a resistivity of material of the i-layer, Θ is a geometrical width of a plate contact (assuming a square shape of this electrode). Measured resistivity ρ depends on a probe position y if material exhibits inhomogeneity of a equilibrium resistivity ρ_0. Resistivity ρ can also be dependent on voltage U due to interface between the needle electrode and the material under test, and due to a finite diffusion length of the injected excess carriers. The LIV induced carriers Δn_0 modify conductivity of the material (with equilibrium carrier density n_0). However, a real probe does not provide an ideal ohmic contact. Thus, density of the injected excess carriers follows a characteristic of the forward biased Shottky barrier: $\Delta n = \kappa \Delta n_0 [exp(eU(t)/kT)-1]$. Summarizing both reasons, ρ dependence on voltage and time (within BELIV current transient) can be expressed as:

$$\rho(U(t), y) = \rho_0(y+a)[1 + (1 + \frac{\mu_p}{\mu_n}) \frac{\kappa \Delta n_0}{n_0} \exp\{-\frac{(r - \mu_p E(r,t)t)^2}{4D_p t} - \frac{t}{\tau_R}\} < \exp(\frac{eU(t)}{k_B T}) - 1 >] =$$
$$\rho_0(y+a)F[r,t,E(r,t),U(t),D,\mu,\tau] \quad .$$

$$(3.2)$$

Here, r is a length of a radius-vector for a definite point within a tested volume of material, $\mu_{n,p}$ are mobilities of carriers, D is carrier diffusion coefficient, τ_R is carrier recombination lifetime, $E(r,t)$ is electric field dependent on time and on spatial coordinate; ρ_0 is an equilibrium resistivity of the material at y point using a probe of a diameter, κ is a dimensionless (adjustable) factor to take into account of the proportionality between carrier density and current. $l_{eff} \sim \int F dr$ is an effective length integrated over the current trajectories, and it should be actually integrated together with $F(r,t,U(t))$ function. The same complication would appear in evaluation of effective surface area S_{eff} for spread currents. Actually, separated simulation of l_{eff} and of S_{eff} is an incorrect procedure, as the currents spreading within a volume under test should be analyzed. Thus integrated (effective) ratio of l_{eff}/S_{eff} (as a single parameter) can be only considered. For high symmetry of the electric field configuration, as in common spreading resistance measurements, the approximation $l_{eff}/S_{eff}=1/2a$ is widely accepted. Actually, a rather low symmetry of the geometrical configuration does not allow simplifications in our case. Therefore, it is a complicated task for the numerical simulations, which also includes solution of a Poisson equation. Numerically simulated distribution of potential and currents for perpendicular geometry of asymmetric electrodes, emulating a configuration of our experimental regime, are illustrated in Figs. 3.1a and 3.1b.

It can be noticed in Figs. 3.1a and 3.1b that concentration of electric field and of potential as well as of current density, respectively, is significantly enhanced at the probe edge. Actually, the strongest field is located at the beginning of a probe. Thus, spatial resolution of profiling can even be higher than that evaluated using a diameter (a) of probe. More attractive way, to evaluate the ratio of l_{eff}/S_{eff}, is a calibration procedure made by varying U_P, τ_{PL} (to find a regime close to the ohmic one) and a set of material samples combined with the definite probe and by attributing the calibrated current value to the peak amplitude of BELIV current transient.

Actually, monitoring of the correlated changes of LIV pulse and of current response enables ones to control the ohmic regime for a single layer. Then, the current measured directly on a load resistor R_L gives a value of $R_{S\perp}$. Subsequently, variations of current values show the changes of $\Delta R_{S,y(i+1)} \sim R_{S,y(i)} l_{eff(i+1)}/l_{eff(i)}$ relative to the previous probe location points.

Interface of a junction is clearly manifested within profiling scan by a crucial change of the BELIV transient shape and of peak amplitude, when needle-probe crosses this junction. The simulated variations of transients dependent on spreading resistance within an elevated resistivity layer are illustrated in Fig. 3.1c, for a needle probe located behind the interface. Barrier charging current is significantly less than an ohmic current within a large conductivity layer (e.g. either metallic electrode or p^+/n^+ layers). Therefore, the peak amplitude crucially drops, while a shape of a transient changes from a LIV-pulse-like to that inherent for charge extraction current (Fig. 3.1c).

Further fragment of the depth-profile of BELIV current transients, recorded relatively to a junction location, provides the additional characteristics for evaluation of $R_{S\perp}$ and for resolution of an impact of the deep centres within high resistivity layer. The influence of the deep traps can be resolved: either i) as an appearance of a recess in the range of barrier capacitance charging peak when the traps filling process at depletion boundary (within λ-layer) ceases moving of this boundary (due to carrier extraction); or ii) as a manifestation of a current minimum within a BELIV current transient with a current increment within the ulterior component of the transient due to an enhancement of generation current, as illustrated in Fig. 2.3.

The increased $R_{S\perp}$ determines a significant delay $(R_{S\perp}C_{b0})$ of an initial BELIV current peak. The barrier capacitance ascribed to the same plate electrode, employed for profiling, could be tested by measuring the BELIV current transient using the parallel-plate electrodes (when $R_{S//}$ can be ignored). Then, the delay time $R_{S\perp}C_{b0}$, obtained from the profiling measurements, enables ones to extract the $R_{S\perp}$ value.

The additional verifications of the electrode capability, as a carrier reservoir, are necessary in such a profiling measurement, to avoid the current limitation due to insufficient quality of electrode. This can be performed by analysis of the amplitude of the current transient dependent on U_P and by combining the reverse and forward LIV biasing regimes.

33

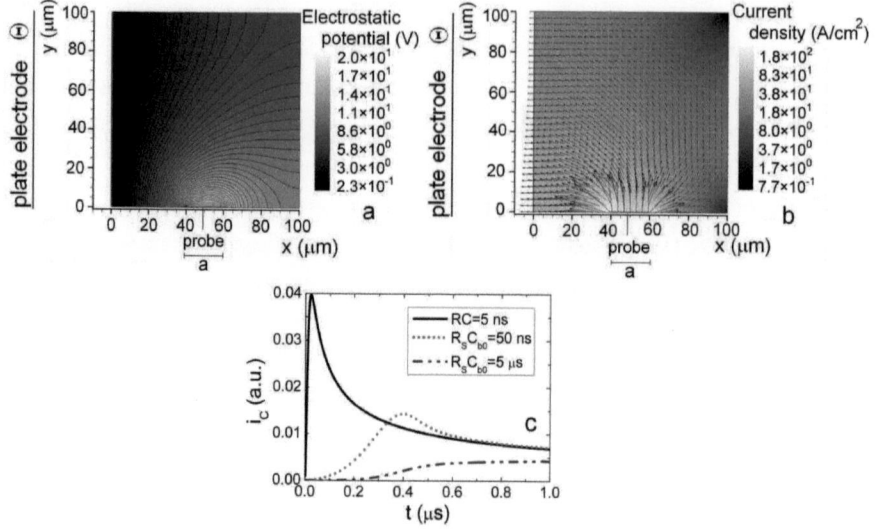

Figure 3.1. Simulated (by using TCAD platform) distribution of potential (a) and of current density (b) for the perpendicularly located probes within a single layer. Values of potential are indicated by a white-gray-black bar in Fig. a. Length of vectors shows simulated current density in Fig. b. c – Comparison of the simulated (at reverse bias) BELIV current $i_C(t)$ transient (for parallel-plate electrodes and for initial delay RC=5 ns - solid grey curve) with those obtained for a needle-tip probe located on boundary of layered structure within an elevated resistivity material layer (behind the interface of the abrupt junction), when delay $R_S C_{b0}$ (broken and black curves) is dependent on spreading resistance $R_{S\perp}$. The broken and black (solid) curves illustrate simulated transients for a BELIV response measured on R_L= 50 Ω using a convolution integral with $R_S C_{b0}$ values of $R_S C_{b0}$=5 ns (black solid), 50 ns (dot), and 5 μs (dash-dot), respectively. (After E. Gaubas et al, ISRN Materials Science, (2012) article ID543790, doi:10.5402/2012/543790, [12]).

The impact of generation current should be controlled by monitoring of the transient shape changes through a varied pulse duration τ_{PL}. The carrier capture characteristics can be examined by using additional steady-state bias illumination to vary filling of traps, in the case if the initial carrier extraction current peak appears to be deformed. The influence of spreading resistance and of serial resistance can be separated by combining the cross-sectional and parallel-plate measurement regimes on the same structure.

The influence of the generation centres in definite layers of junction structure can be better revealed by photo-ionization and thermal-emission spectroscopy measurements. There, techniques of pulsed photo-excitation and bias illumination should be properly combined.

4. Principles of photo-ionization spectroscopy

Photo-ionization or photo-neutralization of deep levels is characterized by optical cross-section, and spectroscopy of these cross-sections provides the direct information concerning matrix elements coupling wave functions of deep levels to those wave functions of band free carriers. Approximation of a potential well ascribed to a definite deep level can commonly be made by either a Coulomb-type or Dirac-type well. Assumption of a δ-type potential of a well leads to the Lucovsky's model [14]. This model is suitable for evaluation of the red threshold of the photoionization energy. However, Lucovsky's model does not incorporate properly the lattice relaxation, Franck-Condon shift and other features of measured spectra. It can be employed to describe the amplitude steps and their red threshold position within a photo-ionization response dependence on the excitation quanta energy ($h\nu$). While the shape of spectral peaks and their shifts can not be simulated relevantly. Nevertheless, Lucovsky's model is a simple approach and has actually only a single parameter E_d of an optical activation energy within a fitting procedure of spectra analysis.

In applications of Lucovsky's approach [14], the photo-ionization of the trapped carriers n_d by using short (fs) pulses of spectrally resolved IR light enables ones to determine the parameters of deep traps and a state of their filling. Short IR pulse with the incident $h\nu$ energy photons of the surface density $F(h\nu)$ integrated per pulse duration makes δ-shape (~ 0.2 ps within scale of BELIV μs pulses) optical emission of trapped carriers, if $h\nu$ fits to E_d and cross-section σ_{p-e} of photon-electron interaction is sufficient. The cross-section σ_{p-e} for electrons located in deep levels is then described [15, 16] by using Lucovsky's expression [14]:

$$\sigma_{p-e}(h\nu) = \frac{BE_d^{1/2}(h\nu - E_d)^{3/2}}{(h\nu)^3}, \qquad (4.1)$$

where B is a multiplicative factor. Change of absorption coefficient $\alpha(h\nu)$ for $h\nu$ light due to photo-ionization of n_{d0} trapped carriers one can describe by

$$\alpha(h\nu) = \sigma_{p-e}(h\nu)n_{d0}.$$
(4.2)

Here, as common, h denotes Plank constant. This change of the absorption coefficient can be controlled by $h\nu$ light induced transmission measurements. IR induced absorption measurements are performed in nearly wave-guide regime within layered device structures. Illumination by a IR light pulse of the surface density $F(h\nu)$ leads to the density of photo-emitted carriers described by:

$$n_d^* = \sigma_{p-e}(h\nu)n_{d0}F(h\nu).$$
(4.3)

By substituting n_d^* (Eq. 4.3) (with $n_d^*=n_d(t))$, into Eq. (4.2), a value of the barrier capacitance charging current modified by the excitation light can be evaluated. Thereby simulated (using Eqs. 2.2 and 4.1 as well as 4.3) BELIV transient (curve 4) is illustrated in Fig. 4.1b. The density of the photo-excited carriers n_d^* by $h\nu$ light pulse can be independently controlled by using the contactless measurements of the microwave probed photoconductivity transients [15]. Then, the density of N_d traps can be extracted using Eq. 4.2. The filling factor n_d/N_d can be controlled by combined measurements of i_C peak value or $\alpha(h\nu)$ (as well as n_d^*) as a function of $F|_{h\nu}$, and the saturation of these characteristics indicates complete photo-ionization of N_d traps. The parameters of the generation current can be extracted by combined analysis of the BELIV current transients measured with and without extrinsic IR illumination. Photo-ionization within an electrically neutral region leads to a reduction of the dielectric relaxation time and of the serial resistance in the junction structure. The enhanced density of excess carriers within ENR decreases a duration (RC) of the initial component representing the rise to peak current and results the shorter Debye length of the transitional layer nearby a depletion boundary.

A more comprehensive model has been developed by A.Chantre, G.Vincent and D.Bois (C-V-B model) [17, 18]. This C-V-B model includes an electron-phonon interaction, which is expressed through a product of the electronic transition probability and the vibrational overlap integral. The cross-section σ_{p-e} of the photon-electron interaction at a photon of energy $h\nu$ is proportional to the matrix element for transition between a deep level d and a band b as

$$\sigma_{p-e}(h\nu) \propto \frac{1}{h\nu}|<b,\vec{k}\,|-i\frac{h}{2\pi}\vec{\nabla}\,|\,\Phi_d>|^2\,\rho(E_k) = \frac{1}{h\nu}M_{bk}\rho(E_k), \qquad (4.4)$$

where energy conservation yields $h\nu = E_0 + E_k$ with E_0 the optical ionization energy including Franck-Condon shift δ_{FC}, and $\rho_b(E_k)$ denotes the number of states with energy E_k in band b represented by a $|b,\mathbf{k}>$ wave function. The electronic wave function of the system with an electron trapped on level d is represented by Φ_d. The latter function can be considered as being built up from the states of band b' and represented by $\Phi_d = \Sigma_k A_k |b',\mathbf{k}>$. The spectral variation of the $\sigma_{p-e}(h\nu)$ is determined by the coefficients A_k of a Fourier transform, by the matrix elements M_{bk} and the density of states ρ_b in the E_k continuum. For the δ-type potential $\Phi_d(r) \propto \exp(-\alpha r)/r$ in Lucovsky's model, one obtains $A_k \propto (k^2 + \alpha^2)^{-1}$. While in A.Chantre, G.Vincent and D.Bois model [17], α controls an extent of a wave function Φ_d, and $\alpha^2 = 8\pi^2 m_b \cdot E_0/h^2$ is related to E_0, and it is derived using the effective- mass m_b approximation. Such an approach yields a distinction between transitions attributed to the photo-ionization ($b=b'$) and the photo-neutralization ($b \neq b'$). The effective mass ascribed to different continuum state bands appears in the definition $k^2 = 8\pi^2 m_b(h\nu - E_0)h^2$ of the wave vectors k. Consideration of the matrix elements M_{bk} yields a forbidden dipole transition for the case ($b=b'$) of photo-ionization $M_{bk} \propto hk$. While these matrix elements $M_{bk} = const$ are independent of hk for the allowed photo-neutralization transitions between states ($b \neq b'$) with different symmetry. On the basis of rather general assumptions, by including a superposition of the elementary components ascribed to transitions to various M_i valleys and by adjustable value P_i of oscillator strength, A.Chantre, G.Vincent and D.Bois [17] derived the total optical cross-section expressed as:

$$\sigma_{p-e}(y) \propto \frac{1}{y\Theta^{1/2}}\sum_i P_i M_i c_i^{2b-a-3/2} \int\limits_{1+\Delta_i/E_0}^{\infty} \frac{[x-1-(\Delta_i/E_0)]^{a+1/2}}{\{[x-1-(\Delta_i/E_0)]+mc_i\}^{2b}}\exp[-\frac{(y-x)^2}{\Theta}]dx.$$

$$(4.5)$$

Here, the symbols represent: it is $a=0$ for an allowed transition and $a=1$ for a forbidden transition; Δi is the energy difference between extrema of the same band relative to the energy minimum point (i.e. Γ); it is $b=1$ for a Dirac type well and $b=2$ for a Coulomb type well; c_i is a ratio of the effective masses $c_i=m^*_\Gamma/m^*_i$; kT is a thermal energy; other parameters are described by relations as $m=h^2\alpha^2/8\pi^2\,m^*_\Gamma E_0$; $y=h\nu/E_0$; $\Theta=4kT\delta_{FC}/E_0^2$. For a single parabolic band, it should be assumed $\Delta i =0$.

However, it should be pointed out that in A.Chantre, G.Vincent and D.Bois model [17] three adjustable parameters (instead of a single adjustable E_d parameter for Lucovsky's model, - if B can be ignored for analysis of a single spectral step) are employed: the band-state wave function extent α^{-1}, the Franck-Condon parameter δ_{FC} ($E_0= E+\delta_{FC}$ with binding energy E), and the relative oscillator strength $P_{i/\Gamma}$. The $\alpha \propto m^{1/2}$ parameter controls the width of a spectral step and the location of its maximum, while the red-threshold energy is nearly unchanged. The $P_{i/\Gamma}$ parameter determines an amplitude of the spectral step. The parameter δ_{FC} characterizes a displacement of the transition threshold energy, while a shape of the spectral curve is almost insensitive to the changes of this δ_{FC} parameter. Nevertheless the A.Chantre, G.Vincent and D.Bois model [17] takes into account a strong electron-phonon interaction, it is more flexible in simulation of the spectral-step shape and the location of spectral maximum. Also, it enables ones to extract a set of the significant physical and spectral parameters.

Usage of the A.Chantre, G.Vincent and D.Bois model [17] is clearly motivated, if photo-ionization and thermal emission responses are simultaneously examined by BELIV technique.

The barrier capacitance $C_b(t) =\varepsilon\varepsilon_0 S/W(t)$ temporal (t) changes under LIV pulse in n-type junction layer are ascribed to variation of a depletion width

$$W(t) =[\frac{2\varepsilon\varepsilon_0 (U_{bi} + At)}{e(N_D +(N_d - n_d(t)))}]^{1/2} .$$
(4.6)

Here ε_0 is the vacuum dielectric constant, ε is the material permittivity, e is the elementary charge, S is the junction area, U_{bi} is the built-in potential barrier, N_D is the shallow dopants density, N_d is the density of deep donor type traps, n_d is the concentration of electrons trapped on N_d, and $A=U_P/\tau_P$ is the ramp of the LIV pulse with amplitude U_P and of duration τ_P. Without additional extrinsic IR illumination, the time dependent changes of the charge $q=C_bU$ within the junction, determine the current transient $i_C(t)$:

$$i_C(t) = \frac{dq}{dt} = \frac{d[C_b(t)U(t)]}{dt} = AC_b(t)[\frac{1+\dfrac{At}{2U_{bi}}}{1+\dfrac{At}{U_{bi}}} + \frac{t}{2\tau_{gd}}\frac{n_d(t)}{N_D+(N_d-n_d(t))}]. \quad (4.7)$$

This transient contains the displacement and conductivity current components. The latter component arises within transitional layer (due to free carrier "tail") at depletion boundary (in more rigorous approach of extended depletion approximation, see [7]) caused by prevailing of carrier thermal generation (from N_d traps) with increase of the depletion width $W(t)$ under reverse bias LIV. Here, it is assumed that trapped (at N_d) carriers are released according to $n_d(t)=n_{d0}\exp(-t/\tau_{gd})$, with steady-state ($N_d$ filling) concentration n_{d0}. The thermal generation lifetime $\tau_{gd}=1/[\sigma_{dth}v_TN_C\exp(-E_d/kT)]$ is a function of the emission cross-section σ_{dth}, of the thermal velocity v_T, of the density of states in the conductivity band N_C, and of activation (E_d) as well as of thermal (kT) energy. As usually, several trap species of different types appear. Then, generation currents (within depletion region) from slower and deeper traps act simultaneously as the leakage current, which is expressed as $i_{gl}(t) =en_iW(t)S/\tau_{gl}$ through the averaged lifetime τ_{gl} as well as through intrinsic carrier density n_i, and added to $i_C(t)$, i.e. $i_\Sigma(t)=i_C(t)+i_{gl}(t)$.

At room temperature, values of carrier thermal generation lifetimes (even for moderately deep ($E_i<E_d<E_C$) centres in Si) appear to be in the range from a few μs to ns (Fig. 4.1a) for traps with σ_{dth} in the range of $10^{-14} - 10^{-16}$ cm^2. Therefore, only traps with short carrier capture lifetimes $\tau_{Cd}/\tau_{gd}<1$ can be filled. Carrier capture lifetime τ_{Cd} is a reciprocal function $\tau_{Cd}=1/\sigma_{dth}v_TN_d$ of the trap density N_d. Thus, the impact of generation current ascribed to the single type traps can be observable as an initial recess within the BELIV transients when density of these traps approaches to or exceeds the concentration N_D of shallow dopants.

The simulated (using Eqs. 2.4 and 4.2) BELIV transients (curves 1-3) obtained varying n_{d0} are illustrated in Fig. 4.1b. It can be noticed in Fig. 4.1b, that an initial recess within the barrier charging current transient is observed when a rather large density of initially filled traps exists. In the opposite case, when several species of traps compete in capturing free carriers, only partial and rather low filling of these traps is possible (carriers of low density are redistributed among different traps). Then, generation-leakage current i_{gl} increases with At voltage and can exceed the barrier charging current in the rearward phase of the transient. The descending charge extraction (displacement current) component and the ascending generation i_{gl} current component imply the existence of a current minimum in the current transient [19], which can be highlighted by increasing a LIV pulse duration.

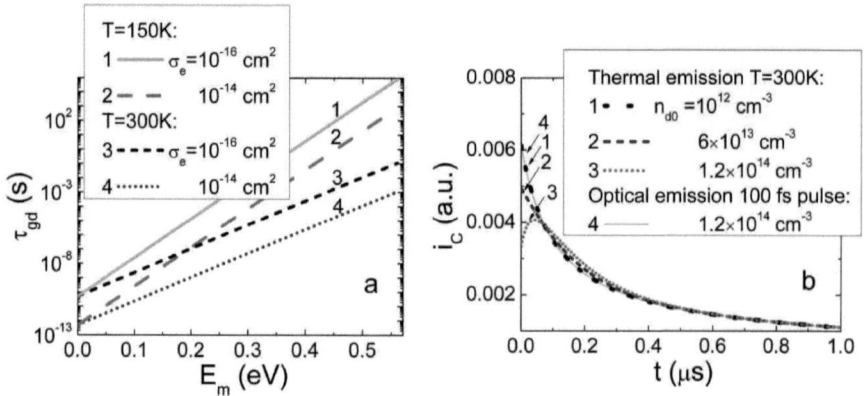

Figure 4.1. a- Simulated thermal emission lifetimes at 150 (1,2) and 300 K (3,4) temperatures for traps with emission cross-section $\sigma_e=10^{-16}$ cm^2 (1,3) and $\sigma_e=10^{-14}$cm^2 (2,4), respectively. b- Simulated barrier ($N_D=7\times10^{13}$ cm^{-3}) charging current transients when carriers of different density n_{d0} trapped at deep donor centres ($N_d=1.2\times10^{14}$ cm^{-3}) are released thermally (1-3 curves) and by short IR light pulse (4). (After E. Gaubas, et al, Journal of INSTrumentation **7** (2012) P01003, doi:10.1088/1748-0221/7/01/P01003, [16]).

The activation energy E_d of N_d traps can be evaluated by spectral measurements of changes in the BELIV current transient shape and the initial peak amplitude, described by Eqs. (4.1-4.5). These spectral measurements are as usually started from the long wavelength wing to avoid simultaneous trap filling/emission from deeper centres. Value of E_d can be estimated as quantum energy $h\nu$, as a red-threshold quantity, for which the IR modified i_C increase is observed. Alternatively, E_d and $\sigma_{pe}(h\nu)$ can be estimated by simulating the steps of the BELIV current amplitudes based on Eqs. 4.1 – 4.5.

The simplified description of principles of the technique of photo-ionization probed barrier capacitance charging transients is presented above by analyzing the donor type traps. Both acceptor and donor type defects acting together can be found in real experimental practice. Analysis of such models (when acceptors behave like the compensating centres, while donors are able to follow fast changes in pulsed voltage) could be found e.g. in monograph [7]. The BELIV-IR (extrinsic infrared) pulsed spectroscopy technique is preferential to evaluate material characteristics at room temperature when density of traps is large [16]. However, an impact of the electron-phonon interactions within room-temperature photo-ionization spectroscopy should be estimated. In high resistivity materials, deep traps of high density are partially filled due to lack of free carriers. Then, generation current prevails in BELIV transients. The mentioned peculiarities are illustrated in Section 5 by results obtained in Si structures containing high density of technological contaminants and radiation defects.

5. Photoionization spectroscopy of deep traps in Si homojunctions

5.1. Material and samples

The BELIV-IR (extrinsic infrared) pulsed spectroscopy technique was applied to analysis of deep level spectra in the Si thyristor and pin detector structures. Industrial non-capsulated Si thyristor n^+pnp structures with well defined layer thicknesses and doping ($N_D = 7 \times 10^{13}$ cm^{-3}, phosphorus doped n-layer) densities were used for recording deep level spectra in a 350 μm thick n-Si layer. The tentative measurements on deep level transient spectroscopy in these thyristor structures showed technological contaminants of significant density within n-layer.

Also, a set of pin diodes of CERN standard p^+nn^+ detector structure (with dopants density of $N_D \cong 10^{12}$ cm^{-3} in n-Si layer of ~300 μm thickness [20]) irradiated by reactor neutrons with fluences in the range of 10^{12} -10^{16} n/cm^2 was investigated. Measurements of current transients were carried out in the range of temperatures from 120 to 300 K. Several types of non-irradiated devices and materials have been also tested in approval of junction evaluation by BELIV technique.

5.2. Illumination dependent barrier capacitance charging transients

For an n-layer containing a high density of deep traps, the transients of the barrier capacitance charging current (BCC) have been measured.

Three type BCC transients have been revealed for p-n barriers in thyristor structures. The **A**- and **C**-type transients, illustrated in Fig 5.1a, contain an initial recess. This recess within a BCC transient indicates a reduced $i_C(t)=AC_b(t)$ barrier charging current due to the trap filling. The **B**-type transients (Fig. 5.1a) are characterized by a restored initial BCC peak. It can be deduced that an illumination pulse of a fixed density and of a fixed wavelength ($\lambda=4$ µm) is able to restore the barrier charging current. This happens when emptying of the carrier capture donor-type centres in the material is saturated by a sufficient density of illumination ($F(h\nu)$). Due to the emptied centres, the initial recess disappears in the BELIV transient. It can also be inferred that the value of the carrier capture lifetime determines the relaxation of the value of the initial current peak associated with the barrier charging (Fig. 5.1a, **B**-type transients). The evolution of BCC transients illustrated in Fig. 5.1a is obtained only for the fixed wavelengths of IR illumination.

Varying the wavelength of this OPO-DFG IR pulsed illumination, a spectrum of deep levels is obtained. Additional investigation performed by measurements of the capacitance voltage (C-V) and current-voltage (I-V) characteristics on the thyristor structures, indicated a large leakage current. These measurements were implemented by separately connecting the different junctions of the structure to a measurement circuitry. The capacitance dependence on the ac test signal frequency suggests that the characteristic lifetimes of carrier capture are of the order of magnitude ~ 1 µs. This is directly corroborated by the analysis of the shape of BCC transients observed for different intensities of the OPO-DFG IR illumination, whereby the generation (leakage) current dominates over the barrier capacitance current for GLIV pulse durations of $\tau_P>1$ µs.

To verify the existence of defects in the n-base region and to identify the traps, responsible for the generation current, C-DLTS spectra have also been recorded on the same device structures. The prevailing peak in the range of 150 -200 K is obtained in the C-DLTS spectrum which is ascribed to the sulphur contaminant impurities [15, 21-23] with activation energy of 0.3 eV and cross-section of $\sigma \sim 5\times10^{-16}$ cm^2. Due to high density of technological contaminants, only a qualitative evaluation of trap concentration by C-DLT spectroscopy is possible when trap density approaches to or exceeds the doping density.

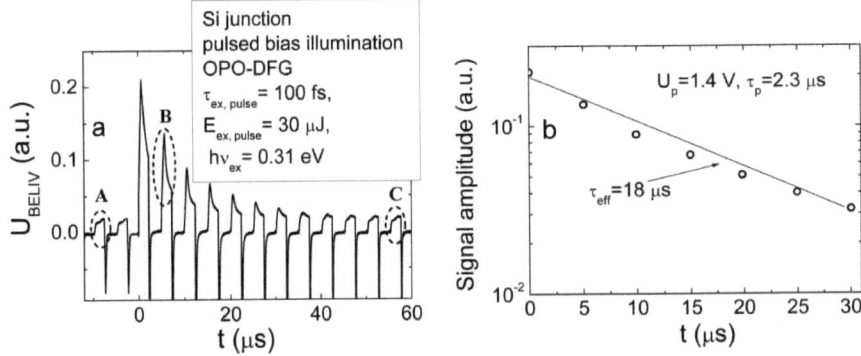

Figure 5.1. a- A sequence of Barrier Charging Current (BCC) transients measured in a Si thyristor junction structure. **A**-type transient indicates a typical transient determined by carrier capture into deep traps before IR illumination pulse, **B**-type transient shows a modified BCC transient just after IR illumination ($\lambda=4$ µm) by 40 fs pulse, and **C**-type transient illustrates the recovered BCC transient after photo-ionized carriers are re-trapped by deep centres. The IR illumination pulse at $t=0$. b- Relaxation of the initial amplitude of the **B**-type transients within semilog scale and evaluation of the carrier recombination/capture lifetime. (After E. Gaubas, et al, Journal of INSTrumentation **7** (2012) P01003, doi:10.1088/1748-0221/7/01/P01003, [16]).

The obtained DLTS spectra actually corroborated the existence of deep traps characterized by activation energy equal to that extracted from the BELIV-IR spectroscopy measurements. This illustrates that the proposed BELIV-IR pulsed spectroscopy technique allows the time resolved spectroscopy with the simultaneous determination of carrier capture and emission lifetimes. Calibration of the illumination density at each wavelength (simultaneously measured in our experiments) enables to determine the concentration of each of the deep traps resolved as a spectral peak in the n-layer of the n^+pnp thyristor structures.

The photo-ionization and thermal emission cross-sections can be estimated by using the directly measured parameters: the ratio r of the amplitudes of photo-capacitance a_{pc} and barrier capacitance a_{bc} determined by dopants, as $r=a_{pc}/a_{bc}$; the doping density N_D; the energy of excitation light E_{op} per pulse and the $F(0.31 \text{ eV})$, i.e.

the surface density of the impinged quanta of a fixed energy, attributed to the spectral peak; the carrier capture time τ_p (Fig. 5.1b). The adjacent pulses in BELIV pulses set can be fitted at a transform point where a B-type transient is changed by a C-type transient. The last pulse of the B-type transient is assumed to be the doping determined barrier capacitance current transient, determined by N_D. Then, the ratio of amplitudes is employed to extract the density excess carriers n_{0ex}, which are photo-emitted from the deep level: $r = n_{0ex}/N_D$. This $n_{0ex} = rN_D = n_{d0}$ (in Eqs. 4.2. and 4.3) is supposed to be the initial filling density n_{d0} of deep level if excitation light intensity is sufficient to photo-emit completely the trapped n_{d0} carriers from the fully filled $n_{d0} = N_T$ level. Thereby using Eq. 4.2, the parameters are related through the absorption coefficient α for a fixed deep level: $\sigma_{e-p} n_{d0} = \alpha = n_{0ex}/F$. The photo-ionization cross-section can then be extracted by using the relation: $\sigma_{e-p} = (rN_D)/(F\, n_{d0})$. Value of $\sigma_{e-p} = 4 \times 10^{-16}$ cm^2 is extracted for $h\nu = 0.31$ eV using the parameters denoted in Fig. 5.1. The cross-section σ_T for thermal emission can be estimated from the measured carrier (with thermal velocity v_T) trapping time $\tau_p = 1/(\sigma_T v_T N_T)$. Assuming that these $N_T = 5 \times 10^{14}$ cm^{-3} traps are filled with excess carriers $n_{0ex} = N_D$ during the last B-type pulse, value of the $\sigma_T = 1/(\tau_p v_T N_T) = 6 \times 10^{-16}$ cm^2 is obtained.

The discussed evaluation is made at assumption of a rather weak electron-lattice interaction, where the process of the photo-neutralization of a deep trap prevails. The obtained parameters are in a rather good agreement with the deep level signatures ascribed to the sulphur contaminant impurities [21-23], - with activation energy of 0.3 eV and cross-section of $\sigma \sim 5 \times 10^{-16}$ cm^2 signatures, extracted from the C-DLTS spectra in the range of 150 -200 K.

5.3. Illumination spectrum dependent BELIV transients in the irradiated diodes

Prevailing of the barrier charging current (Fig. 5.2a) has been observed in Si pin detectors irradiated with a rather low fluence of reactor neutrons ($\Phi \leq 10^{13}$ n/cm^2). The photo-ionization peaks, probed by the BELIV current measurements, are determined varying illumination quanta $h\nu$. These peaks simulated by using Eqs. 4.1- 4.5 (and $[N_D - (N_{AT} - n_T(t))]$ when necessary, with N_{AT} the density of acceptor-type traps) can be associated with well-known [24] deep centres ascribed to radiation defects.

The evolution of the BCC transients, illustrated in figure 5.2a, as a function of the illumination wavelength for the same quanta flux (calibrated OPO-DFG energy per pulse) shows the photo-ionization peaks with activation energy values of $E_1=0.3\pm0.02$, $E_2=0.41\pm0.01$ and $E_3=0.51\pm0.01$ eV in an n-Si layer with a dopant density of $N_D\approx10^{12}$ cm^{-3} and irradiated with $\Phi=10^{13}$ n/cm^2. In Fig. 5.2a, it can also be noticed, that the capture lifetime varies for different deep traps, as deduced from the BCC amplitude reduction rate within a sequence of transients at fixed illumination wavelength.

In Si pin detectors irradiated with $\Phi \geq 10^{14}$ n/cm^2 fluence, the generation current i_{gl} dominates in the BCC transients when the deep trap density exceeds that of the shallow dopants. This is held even using the highest illumination density of the OPO-DFG source, being the IR spectrum brightest laser, and rather short LIV pulses [16, 25]. The evolution of the BCC transients shown in Fig. 5.2b can be explained by a simultaneous increase of the generation current (i_{gl}) within space charge region and the dielectric relaxation time. This leads to an increase of the series resistance of the ENR n-Si region. A high density of deep traps also leads to a partial filling of the different deep centres due to a lack of "native" equilibrium carriers which is determined by the shallow dopants of low ($N_D\approx10^{12}$ cm^{-3}) concentration. Then, deep traps are characterized by the longest carrier emission lifetime (i.e. the deepest centres) which exceeds the carrier capture lifetime can be filled [7].

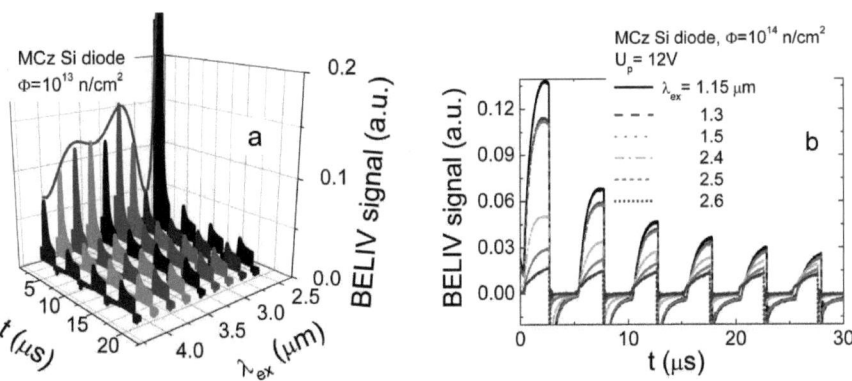

Figure 5.2. Illumination spectrum dependent variations of the BCC transients measured in Si pin detector structures irradiated with 10^{13} (a) and 10^{14} (b) n/cm^2 reactor neutrons fluence. (After E. Gaubas, et al, Journal of INSTrumentation **7** (2012) P01003, doi:10.1088/1748-0221/7/01/P01003, [16]).

This prediction is proven in our experiments on the Si pin detectors, heavily irradiated with neutron fluences in the range of $\Phi=10^{14}$ to 10^{16} n/cm^2. There a threshold for the photo-ionization is observed in the range of quanta energy $h\nu \geq 0.5$ eV. The BELIV-IR spectral peak revealed in Fig. 5.2b can only be obtained for an OPO illumination wavelength of $\lambda \cong 1.2$ μm, which is close to the inter-band excess carrier photo-generation. This illustrates that the BELIV-IR spectroscopy technique is applicable even in the case of a high density of traps.

One of the limiting factors in extraction precision for the activation energy (when using threshold wavelength for photo-ionization of definite traps) is the spectral width $\Delta(h\nu)$ of the OPO radiation. The spectral broadening of fs pulsed radiation comprises about $\Delta(h\nu)$ =10-80 meV, and this broadening increases with wavelength. However, at room temperature, relatively shallow levels are thermally ionized, and it comprises $\Delta(h\nu)/h\nu \cong$ 2-20 % for the observed peaks. OPO pulses of ps duration are preferential to reduce spectral broadening of tuneable wavelength IR illumination. Actually, illumination at peak wavelength dominates within the photo-ionization of trapped carriers when using fs pulses. The BELIV-IR spectroscopy technique is preferential when traditional methods are non-operational (e.g. TSC – thermally stimulated currents at 300 K and DLTS at high density of several species traps in samples with large leakage currents).

In summary, the proposed BELIV-IR pulsed spectroscopy technique [11-13, 15, 16, 19, 25] can be used as a powerful tool for the examination of deep levels in the junction area of semiconductor device structures even when a high density of defects is present. This technique allows simultaneous measurements of the excess carrier capture and short emission lifetime in the resistive junction layer of a semiconductor device structure. Comparison of spectra measured by the BELIV-IR pulsed with those registered by C-DLTS on the same Si thyristor structures shows an excellent agreement between both techniques. This is valid even in the case where the dopant density is close to that of the trap concentration. One of the additional advantages of the BELIV-IR technique is that it can be performed at room temperature. Such a measurement regime is closer to the real operation of the devices, and it can be implemented without the need for temperature scanning in a cryostat.

6. Barrier evaluation for heterojunctions

The BELIV technique can also be a suitable tool for the characterization of the poly-crystalline heterojunctions. Actually, processes in the large resistivity layer are controlled under depletion regime. Nevertheless, interplay of defects can be deduced from the signatures determined by BELIV technique in the base region of a junction. The applications of the BELIV technique for examination of heterojunctions are illustrated using of the polycrystalline films containing Cu_2S-CdS heterostructures.

6.1. Samples investigated

The polycrystalline films containing Cu_2S-CdS heterostructures formed by dry deposition method [26] have been investigated. The heterostructures were formed by employing a substitution technique where a layer of copper sulfide is formed directly on the substrate layer of CdS during heat treatment using a pre-printed cuprous chloride film, formed by vacuum evaporation. However, such a heterojunction consisting of p-type Cu_2S and n-type CdS is a rather complicated heterostructure because of the interface between two materials with different electron affinities, band gaps, and polycrystalline structures. The lattice mismatch and inter-diffusion of components causes defects at or near the interface that strongly affect the junction properties. In particular, it is shown that copper diffusion into the adjacent CdS layer changes stoichiometry of the Cu_xS layer and might form shunting channels within the CdS layer [27]. Due to polycrystalline structure and lattice mismatch between the adjacent layers, a large density of trapping and recombination centres occurs. These disordered material areas cause the effect of photo-induced modulation of junction potential barrier. To evaluate the impact of traps on parameters and stability of heterojunctions, the barrier capacitance characteristics have been studied by BELIV technique [28].

6.2. Barrier capacitance characteristics

The blocking junction of Cu_2S-CdS for majority carriers has been qualitatively tested by varying polarity of the applied voltage and by measuring I-V, C-V and BELIV signals. I-V asymmetry (Fig. 6.1a) and the shape of the reverse biased BELIV transients (Fig. 6.1b) indicate that a high resistivity layer exists, which is ascribed to the n-type conductivity CdS material. The initial component due to barrier capacitance charging current rise (Fig. 6.2) is an inherent feature for the BELIV pulse of the reverse biased junction.

Three inherent types of BELIV transients, associated with CdS polycrystalline layer properties, have been unveiled on several sets of the investigated in dark Cu_2S-CdS structures. The square-wave shape BELIV transients (Fig. 6.1b, solid black curve) are inherent for the type-I samples. This type of the BELIV transients implies the insulator-like base material with low density of free carriers. The BELIV transients (shown in Fig. 6.1b by a dotted grey curve) with the inherent component of the generation current is pronounced within the rearward wing of the transient which caused by carrier emission from the deep traps in the samples of type-II. Also, the fast increase of this current component implies non-linearity of the rearward contact in the samples of the III-type, appearing at elevated voltages. It can be deduced from the recorded BELIV transients for the samples of the III-type. For non-perfect junctions the rise to peak appears to be smoothed by the large τ_{RC} of the circuit.

For the properly operating junction, the initial current rise to peak is followed by the descending current component due to a reduction of barrier capacitance caused by the charge-extraction (Figs. 6.2b and 6.2c) when a depletion width increases with an enhancement of a voltage during LIV pulse. However, at the presence of the large density of generation centres, the current commonly increases in the rearward component of the transient during evolution of a LIV voltage pulse.

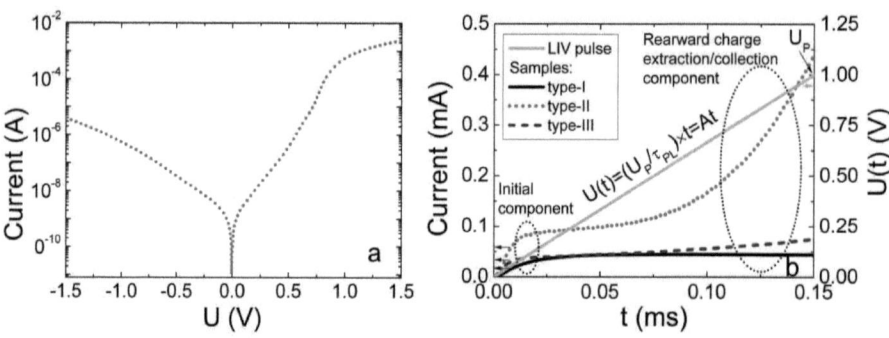

Figure 6.1. The typical I-V (a) and BELIV (b) characteristics observed in Cu_2S-CdS junction structures of type-I (solid black curve), of type-II (dotted grey curve), and of type-III (dark grey dashed curve). Linearly increasing voltage pulse is also presented (light grey). The components of BELIV currents are denoted in Fig. 6.1b by dot circles. (After E. Gaubas et al, Thin Solid Films, **531** (2013) 131, doi:10.1016/j.tsf.2013.01.010, [28]).

For the type-III samples (Fig. 6.1b, dark grey dashed curve), the peak ascribed to the barrier charging current initially prevails. It is superimposed by the ascending generation current component within rearward wing of a transient. The observed differences among the BELIV transient shapes can be explained assuming variation in the relative densities of the dopants, forming shallow levels, and carrier capture centres, associated with deep levels. For the samples of type-I, the density of dopants is small. Therefore, the high density of traps is sufficient to capture nearly all the free carriers. This leads to an insulating state of CdS material, where the base region is fully depleted. This determines the capacitor-like behaviour of the Cu_2S-CdS junction structure under varied reverse voltage.

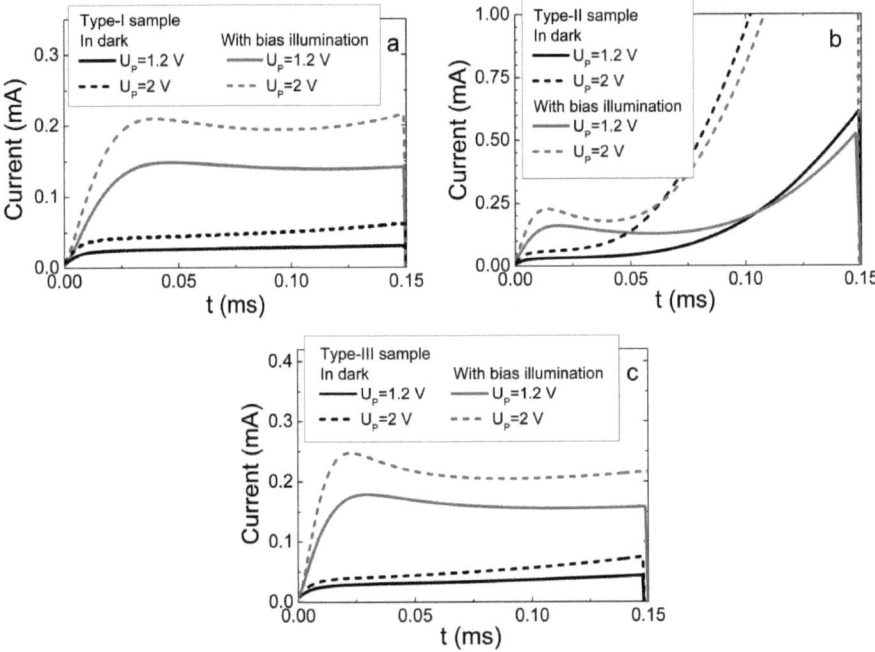

Figure 6.2. Comparison of experimental BELIV transients for different pulsed reverse bias voltage, measured in dark (black curves) and under constant bias illumination intensity (grey curves), recorded on samples of type-I (a), type-II (b), and type-III (c). (After E.Gaubas et al, Thin Solid Films, **531** (2013) 131, doi:10.1016/j.tsf.2013.01.010, [28]).

For the samples of type-II, the density of the shallow impurities is large enough to avoid the full-depletion state within the 20 μm thick CdS layer. However, the density of carrier traps in these type-II samples is also large. Thereby, the generation current component prevails within the rearward component of the BELIV transient in these samples of type-II.

In the samples of type-III, both the barrier charging and the charge extraction currents are present within BELIV transients. These BELIV transient components indicate that the total density of dopants is significantly larger than that of traps.

The inherent features of junctions can additionally be highlighted by measuring the BELIV transients and by varying the steady-state bias illumination. For the type-I samples with the largest resistivity of CdS layer, the bias illumination is insufficient to recover the barrier capacitance (Fig. 6.2a). Therefore, full depletion persists in this CdS layer. The photogenerated carriers in the samples of type-II (Fig. 6.2b) and type-III (Fig. 6.2c) lead to the formation of a pronounced peak ascribed to the barrier capacitance charging current. In the samples of type-II, the ascending component of the generation current is shifted to the rearward side of a BELIV pulse. These observations can be understood by the presence of the different density of traps and of the dopants in these various type samples. In particular, type-I samples contain the smallest density of dopants and the largest density of traps. The density of dopants is increased in type-II and type-III samples, while concentration of traps is larger in the samples of type-II.

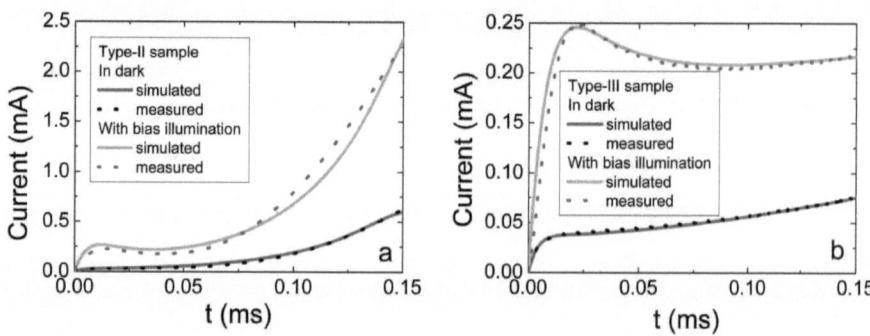

Figure 6.3. Fitting of BELIV transients measured (dotted curves) in dark (black curves) and under illumination (grey curves) for samples of type-II (a) and type-III (b) with the corresponding simulated transients (solid curves). (After E. Gaubas et al, Thin Solid Films, **531** (2013) 131, doi:10.1016/j.tsf.2013.01.010, [28]).

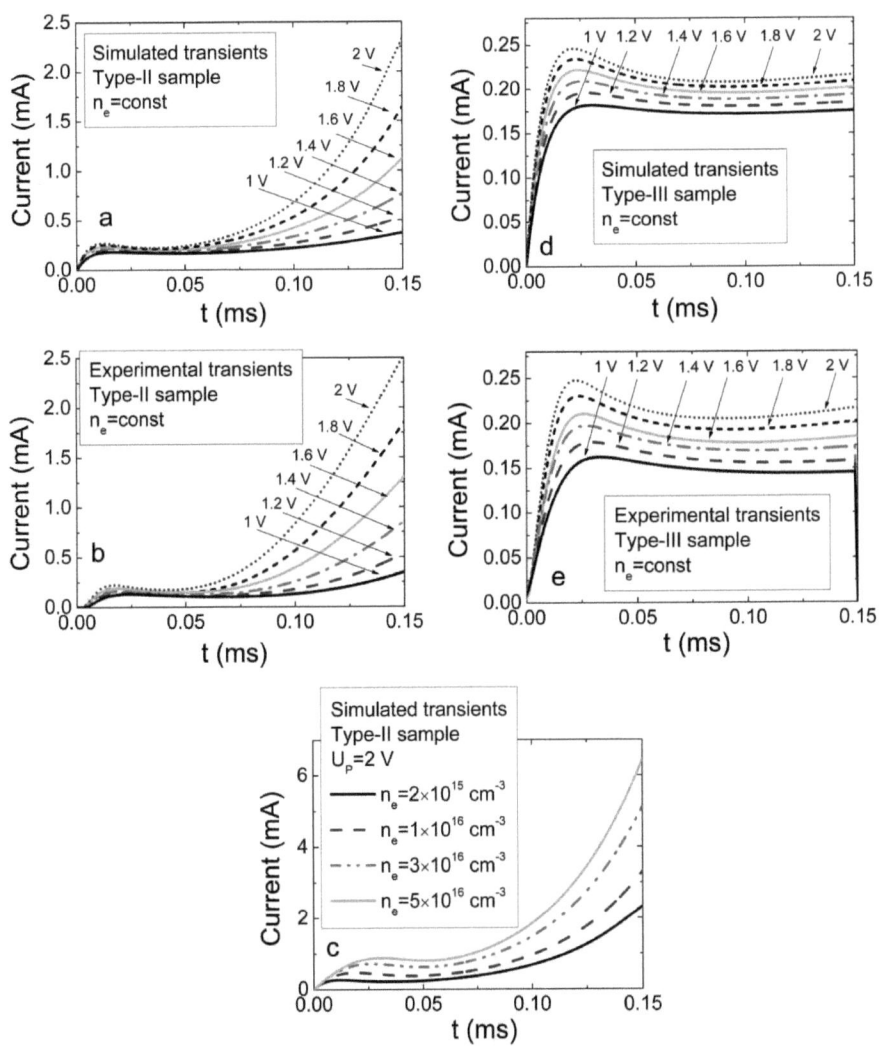

Figure 6.4. Simulated evolution of the BELIV transients for type-II (a and c) and type-III (d) samples compared with the experimental BELIV transients, dependent on U_P, and recorded on type-II (b) and type-III (e) samples, varying U_P and keeping constant density of photogenerated carriers. Simulated transients for type-II sample (at constant $U_P = 2$ V and varying density of photogenerated carriers) are also presented in Fig. 6.4c. (After E. Gaubas et al, Thin Solid Films, **531** (2013) 131, doi:10.1016/j.tsf.2013.01.010, [28]).

The time scale of LIV pulses, over which the characteristic changes of the BELIV transients can be reliably recorded, represents the order of magnitude of the dielectric relaxation time τ_M. This time is directly determined by the density of equilibrium free carriers n_0 and, consequently, by the dopant concentration. The dielectric relaxation time is commonly expressed by relation $\tau_M = \varepsilon\varepsilon_0/en_0\mu$. It can be simply evaluated provided that the mobility μ of majority carrier and the permittivity ε of material are known.

The doping value of $n_0 = 1.5\cdot10^{13}$ cm^{-3} has been estimated from the measurements performed by varying LIV pulse duration. Value of the built-in potential barrier U_{bi} can then be extracted using the measured value of the current peak $i_C(0) = AC_{b0}$ caused by the barrier charging. The built-in potential barrier U_{bi} is consequently evaluated using a relation $C_{b0} = \varepsilon\varepsilon_0 S/w_0 = (\varepsilon\varepsilon_0 S^2 eN_{Def}/2U_{bi})^{1/2}$. Value of $n_0 = N_{Def}$ can independently be estimated by measuring the duration of the current rise to its peak τ_{RC}. The $\tau_{RC} = C_{b0}R_{ENR}$ value can be calculated by combining values of the barrier capacitance C_{b0} and the serial resistance R of the electrically neutral region (ENR) $R_{ENR} = (1/en_0\mu)(d - w_0)/S$. Here, d is the thickness of a junction base layer.

The procedures of the parameter extraction discussed above should be assumed as the first iteration for the more precise evaluation of a batch of the junction parameters. The more precise and comprehensive extraction of values of the junction parameters is implemented by simulating the BELIV transients and by fitting them with measured ones.

In order to get the comprehensive correlation of the simulated and fitted transients, a set of BELIV transients has been measured by varying the reverse LIV voltage values U_P. This set of BELIV transients has been simultaneously simulated, keeping the same set of junction parameters. The simulated transients and their fit to experimental ones are illustrated in Fig. 6.3, as performed for the type-II and type-III samples.

The generation current component is clearly resolved for the type-II samples, and a set of parameters $n_0 = 1.5\cdot10^{13}$ cm^{-3}, $U_{bi} = 0.7$ V and $\tau_g = 6.5$ µs has been extracted using the described fitting procedures (Fig. 6.3a). The non-linearity of the electrodes has been revealed in the range of the elevated voltage values within the rearward component of BELIV transient for the type-II samples. In the same way, a set of parameters $n_0 = 1.5\cdot10^{13}$ cm^{-3}, $U_{bi} = 0.7$ V, $\mu = 0.9$ cm^2/Vs has been determined for the type-III samples. For type-I samples, a geometrical capacitance C_g can only be evaluated. It was found to be equal to 3 nF, and this C_g value is in agreement with that calculated using independently measured thickness d and probed area S of the CdS layer. To extract the set of parameters, the best fit between experimental and simulated transients has simultaneously been achieved for several transients (shown in Figs. 6.3a and 6.3b).

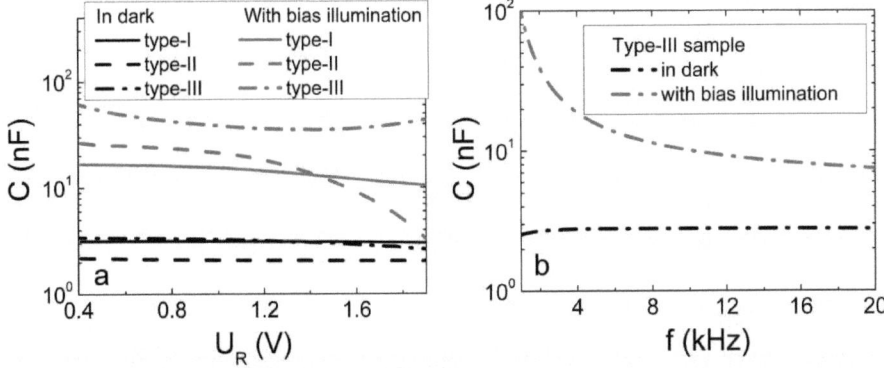

Figure 6.5. a- Experimental C-V characteristics in type-I (solid curves), type-II (dashed curve), and type-III (dash-dotted curve) samples measured using serial connection of junction structure to LRC-meter and recorded without (black curves) and with (grey curves) bias illumination. b- Barrier capacitance dependence on test signal frequency obtained for type-III sample without (black curve) and with (grey curve) bias illumination measured using serial connection regime. (After E. Gaubas et al, Thin Solid Films, **531** (2013) 131, doi:10.1016/j.tsf.2013.01.010, [28]).

In order to verify the characteristics of the junction and to manipulate a filling/emptying of traps, the BELIV transients have been examined using the broad-spectrum illumination (Fig. 6.2). Modifications of BELIV transients under additional illumination and LIV pulse parameters are illustrated in Fig. 6.4. These modifications are caused by the changes of the barrier capacitance, the generation current and the serial resistance. It has been obtained that the barrier capacitance $C_{b0}(N_{Def})=\varepsilon\varepsilon_0 S/w_0(N_{Def})=(\varepsilon\varepsilon_0 S^2 e N_{Def}/2U_{bi}(N_{Def}))^{1/2}$ varies with excess carrier density n_e. Here, $w_0(N_{Def})=(2\varepsilon\varepsilon_0 U_{bi}(N_{Def})/eN_{Def})^{1/2}$. This leads to the light induced modification of the effective doping density $N_{Def}(n_e) = \Sigma_{kj}(N_{Dk}(n_e)-N_{Aj}(n_e))$. Variation of the instantaneous value of the effective doping appears due to photo-emission of carriers from deep acceptors and donors. The generation current changes due to $w(N_{Def})$ and photo-/thermo-emission from traps, as

$i_g(N_{Def})=[en_iSw(N_{Def})/\tau_{gi}+\Sigma_{kj}en_{e,kj}Sw(N_{Def})/\tau_{ge,kj}]$. The photo-generated excess carriers (of density n_e) also modify the serial resistance R_{ENR} (and consequently τ_{RC}), which depends on the total free carrier density (n_0+n_e).

The simulated BELIV transients obtained by varying of LIV voltage U_P (at fixed bias illumination, Figs. 6.4a, 6.4d) and by changing the excess carrier density (at fixed LIV pulse, Fig. 6.4c) are compared with the experimental (Figs. 6.4b, 6.4e) ones in Fig. 6.4. The simulated evolution of transients fairly reproduces the changes of BELIV current pulses, experimentally measured on samples of type- II and type - III (Fig. 6.4).

Variations of the components of barrier capacitance and generation current observed within the BELIV current transients are corroborated by changes of C-V characteristics (Fig. 6.5a) measured on samples of different types. Irrespective of the sample type, the C-V characteristics measured in dark show the capacitance values nearly independent of the applied reverse voltage. Such a C-V behaviour implies a full depletion of the junction base region and, consequently, low density of free carriers. Steady-state illumination determines photo-ionization of deep traps and, thus, the enhancement of free carrier density within a neutral region. This leads to an increase of the barrier capacitance value due to the enhancement of effective doping $C_{b0}(N_{Def}) \sim N_{def}^{1/2}$. This result qualitatively correlates with changes of Cu_2S-CdS junction characteristics under illumination, published in [27]. However in our experiments, the changes of the barrier capacitance have been obtained to be different (Fig. 6.5a) for various types of samples. For the type-I samples, the barrier capacitance is nearly independent of voltage. Therefore such junctions behave like a capacitor due to a lack of free carriers. The changes of BELIV transients in the photo-biased samples of type-II and type-III indicate a recovery of the junction under steady-state illumination. Specifically for type-III samples, barrier capacitance starts to increase at elevated voltages due to non-ohmic contacts and generation/leakage currents. Thus, all the investigated junction types exhibit a wide spectrum of defects acting as carrier capture/generation centres. Presence of high density of the carrier generation centres is additionally confirmed by differences in C-V characteristics obtained for serial (C_S) and parallel (C_P) measurement regimes. There a generation/leakage current prevails (exceeds the test signal current using LRC technique) in C_P characteristics recorded in the range of rather low frequencies and voltages [25]. Improvement of the junction operational features under steady-state broad-spectral-band illumination is also observed in barrier capacitance dependence on test signal frequency (Fig. 6.5b). Bias illumination improves the operational characteristics of a junction. Barrier capacitance (C_b) is nearly frequency (f) independent for junctions kept in dark. This observation implies a capacitor-like (or fully depleted base) qualitative behaviour of the non-photo-biased junction. For the additionally illuminated junctions, the C_b parameter decreases with enhancement of frequency. Such a characteristic is inherent for the partially depleted junction, where dielectric relaxation and carrier capture/emission times determine a rapidness of the junction response. This is an indication that the relatively deep levels determine a frequency range of the resolvable variations within C-V and C-f characteristics. The revealed differences in doping and trap densities, which lead to the significant variations of the electrical characteristics of junctions can additionally be modified by the layer deposition regimes influencing the layer thickness, composition, and structure. Substrate temperature and layer deposition duration were the main parameters varied during the heterojunction formation process. The thickness of the base layer defines its resistivity and the size of the formed microcrystals. Samples with the thicker base CdS layer showed capacitance transients specific for the type-I samples. The separated types of samples (characterized by the inherent capacitance transients) also correlate with Cu_xS layer thickness. The longer is the film deposition time, the closer the C-V and BELIV characteristics are to those inherent in the type-II samples.

The impurities diffused into CdS layer can serve for formation of the acceptor centres that compensate or even overcompensate the donors initially existing in the CdS layer [26]. Moreover, Cu diffusion between the grain boundaries of CdS microcrystals can cause the shunting effect. The conditions for Cu diffusion are directly determined by the deposition time t_{dep} and the substrate temperature T_{sb}, thereby resulting in variations of the electrical characteristics of junction structures. The study of the Cu_2S-CdS junctions of polycrystalline material layers by combining the current-voltage (I-V), the capacitance-voltage (C-V) and the barrier evaluation by linearly increasing pulsed voltage (BELIV) techniques enabled us to reveal and characterize several types of samples. The separated types of heterojunction structures, based on the barrier capacitance characteristics of the junctions, correlate qualitatively with differences in structural composition of the polycrystalline layers, revealed by XRD technique. The enhancement of the doping density and a reduction of the deep centre concentration by manipulating the layer deposition regimes would enable a fabrication of Cu_2S-CdS junction structures with appropriate electrical characteristics.

7. Fluence, temperature and external steady-state bias dependent BELIV characteristics for the irradiated Si diodes

Potential barrier in particle detectors, implemented by a pin structure formation, and its stability under irradiations are the essential characteristics of particle detectors [25]. In such detectors heavily irradiated by high energy particles, formation of the extended defects within a range of metallurgic junction is very probable [24, 29]. Then junction interface circuitry shortening by micro-plasma generation and other junction damage effects can appear. Similarly defects introduced by technological (for instance, doping) procedures are responsible for emerging of a deep level system within a band-gap of semiconducting materials. The deep levels aggravate the parameters of a junction and its operational characteristics. Scanning of the depth distribution of spreading resistance (R_S) is a common and widespread tool for evaluation of the doping profiles within layered structures [6, 30-32]. This method is well standardized for scanning spreading resistance microscopy (e.g. [6]), approaching to spatial resolution of a few nm [32]. It is often used for the technological control in formation of the device structures. However, doping and steady-state carrier density profiles can be insufficient to evaluate and to predict the operational characteristics of fabricated junctions if additional defects are introduced together with dopants.

The small test signal technique implemented by the LCR measurements, to determine a phase shift between reference test voltage signal and a current, which appears due to barrier capacitance charging, is widely applied for a weakly irradiated device. However, these LCR measurements provide the correct barrier capacitance response if a displacement current dominates. The elevated generation current component appears due to thermal emission of trapped carriers in heavily irradiated materials. The generation current is able to completely hide the displacement current if density of traps is close or even exceeds the density of shallow dopants. The generation current leads to a rapid (in terms of low frequency) dissipation component, which diminishes the quality factor [6] of LRC-meter to the unacceptable values.

In the range of low frequencies $1/\tau_{gen} \leq \omega \ll 1/\tau_{NDef}$, the generation current temperature changes can be applied for spectroscopy of deep levels by applying a well-known C-DLTS [6, 7, 33] technique. However, validity of the DLTS usage is held if trap density N_T is sufficiently small in respect to N_{Def}. In the case of $N_T > N_{Def}$, a potential barrier is completely determined by N_T due to requirement of charge conservation. Then, barrier charging becomes rather slow: $\tau_{gen} \sim \exp(E_T/kT)$ for deep traps ($E_T \gg kT$) with activation energy E_T relative to a thermal energy kT. Therefore, development of applications of the comprehensive techniques for testing of junction device structures are desirable, and BELIV technique can be a powerful tool for such examination. A linear increment of the bias voltage is beneficial to reach the small changes of the depletion width with time. The latter technique is also preferential to have a constant ramp of the external voltage variation and of the electric field induced convection and displacement currents. Small and monotonic changes of depletion width are preferential to register the characteristic times of the fast thermal generation processes attributed to the rather shallow centres in the range of operation temperatures of devices under test and to identify the full depletion regime for heavily irradiated diodes. This technique has been applied to analysis of barrier quality of Si detectors irradiated by reactor neutrons (1 MeV eq.) and protons with fluences in the range of 10^{12} - 3×10^{16} cm^{-2}.

7.1. Samples

Particle pad-detectors, fabricated from n- type CZ (Czochralski grown) and MCZ (Czochralski with applied magnetic field grown) Si supplied by Okmetic Ltd have been studied. The non-irradiated and irradiated with reactor neutrons to fluences in the range of $\Phi = 10^{14}$-10^{16} n/cm^2 devices were examined. The detectors with (p$^+$-n-n$^+$) diode structure had an active area of 5×5 mm^2 which was surrounded by 16 floating guard rings. The diode base material had dopants density of about 10^{12} cm^{-3} and base thickness 300 μm. A more detailed description of the detector structure is provided in Ref. [12, 34].

7.2. Fluence and temperature dependent BELIV characteristics

Variations of the charge extraction and injection BELIV currents as a function of voltage increase within LIV pulse, are illustrated in Fig. 7.1. These characteristics show a clear dependence on neutron irradiation fluence.

For the reverse bias LIV pulses, it is clearly observed that the charge extraction current dominates at the relatively small fluences. This enables one to evaluate the barrier parameters, such as U_{bi} and N_{Def}, by employing different regimes of BELIV technique. The N_{Def} seems to be the dominated factor, and C_{bo} decreases with enhancement of fluence from 10^{12} n/cm^2 to 10^{16} n/cm^2, Fig. 7.1. This can be explained by increment of density of the compensation centres, located within a lower half of the band-gap. This increment of density of the compensation centres enhances the initial depletion width w_0 and, subsequently, decreases value of the C_{bo}. However, starting from fluences of 10^{14} n/cm^2, the generation current becomes the prevailing component of a transient at the reverse bias voltage. Thereby, the BELIV current transient exhibits a trapezoid shape starting from the initial step in diodes irradiated with fluence of 10^{14} n/cm^2. Close values of C_{bo} and the trapezoid shape of a transient are obtained for fluence values in the range of 10^{14} cm^{-2}. The generation current also determines a pedestal within a complete current of BELIV transient, and it compensates a reduction of C_{bo}. Therefore, a comparison of the initial step for the reverse and forward biased diode should be employed to exclude this pedestal.

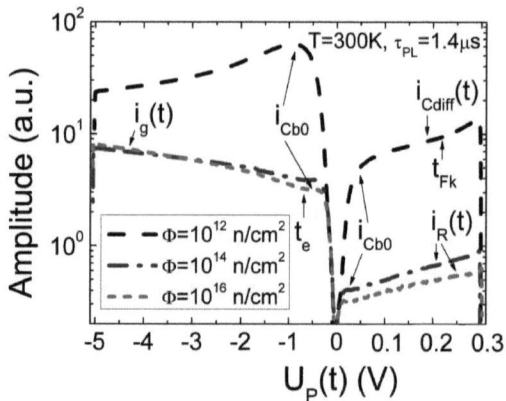

Figure 7.1. Variations of barrier and diffusion capacitance charging currents dependent on irradiation fluence as a function of voltage increase within LIV pulse in Si pin detector, at the same LIV parameters. (After E. Gaubas et al, ISRN Materials Science, (2012) article ID543790, doi:10.5402/2012/543790, [12]).

The initial step in BELIV current transients for the forward biased diode (when C_{bo} prevails) should coincide with that for a reverse biased diode. It is clear from the comparison of the transients that C_{bo} values are of close magnitude for 10^{14} n/cm^2 irradiated diodes (including a difference in U_P ramp). It can be noticed, that in a reverse biased diode, generation current increases with fluence Φ, and, at $\Phi=10^{16}$ n/cm^2, it exceeds that value measured for the 10^{14} n/cm^2 irradiated diode. The amplitude of the initial step also decreases for the forward biased diode. This correlates well with results obtained for the reverse biased diode. The mentioned variations of the BELIV characteristics with irradiation fluence can be explained by approaching of value of the initial barrier capacitance to that of the geometrical capacitance. It is worth noting that the impact of the storage capacitance can be observed only in lightly (10^{12} n/cm^2) irradiated diode at a rather small forward biasing voltage.

For the forward biased diode (a right side characteristic in Fig 7.1), the barrier capacitance prevails in the initial stages of the BELIV transient measured on a diode irradiated with relatively small fluence. Then reduction of stored charge and of diffusion length of the injected carriers leads to a decrease of forward current within a transient, if carrier lifetime becomes shorter than that of a LIV pulse duration. This result is in excellent agreement with carrier lifetime values measured directly using the microwave probed photo-conductivity transients (MW-PCD) [15] for the same samples. The reduction of C_{bo} is determined by an enhancement of the density of compensating centres, and it is reduced to C_{geom} if the equilibrium depletion width approaches to a sample thickness.

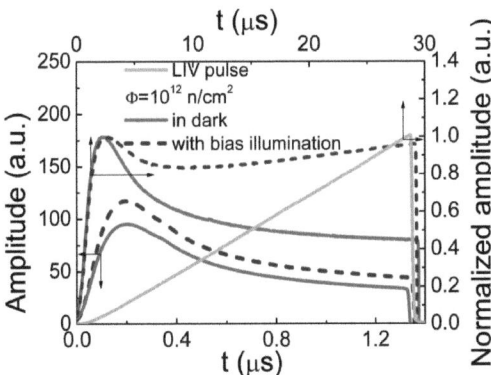

Figure 7.2. Bias illumination (BI) dependent charge extraction current transients (as measured – solid curves, and normalized to a peak amplitude – broken curves) measured on the same irradiated diode at a fixed reverse (U_R) voltage of LIV pulses. (After E.Gaubas et al, ISRN Materials Science, (2012) article ID543790, doi:10.5402/2012/543790, [12]).

For the non-capsulated diode structures, trap determined modifications of the BELIV current $i_C(t)$ transients can be suppressed through the priming of the trap filling by infra-red (IR) continuous wave illumination. Additionally, the initial component of BELIV current transient can be manipulated by a pedestal (by shifting position of Fermi level) of varied polarity of the dc voltage together with a LIV pulse. Variations of the mentioned external factors can be combined with temperature variations to modify filling of the traps.

In Fig. 7.2, variations of the BELIV current transients measured with and without bias illumination (BI) are presented. The primary steady-state illumination modifies the initial filling of traps for electrons and holes. It has been revealed in our investigations on spectral efficiency of bias illumination that spectral range of the inter-band absorption (1.12 eV for Si at room temperature) is optimal for primary filling of traps by IR light. Depending on duration of LIV pulses and on density of traps, the time scale of carrier generation processes can be highlighted, as illustrated in Fig. 7.2. The absolute value of the initial amplitude of BELIV current transient for a reverse biased diode increases proportionally to the intensity of the bias illumination. Simultaneously, a component of the generation current within ulterior stages of a transient is increased with BI. The increment of generation current is more obvious for the normalized (to the amplitude of the initial step) transients, shown in Fig. 7.2. This result clearly proves that the generation current is caused by several traps characterized by a wide spectrum of levels within the upper half of a band-gap, as simulated in Fig. 4.1b. Calibration measurements of the density of the absorbed BI quanta enable ones to evaluate the density of capture centres, for which the saturation of the amplitude of BELIV current is reached, like in photo-ionization spectroscopy, - Section 5.2. For a rather long LIV pulse, duration of which corresponds to the thermal emission time scale (characteristic for a definite sample), the generation current becomes dominant within the rearward component of a transient. BELIV current transients for the reverse bias LIV pulses can be manipulated by the external light only in diodes irradiated by the rather moderate fluences, $\leq 10^{13}$ n/cm^2, at room temperature.

Complementarily, filling of the minority carrier trap can be implemented by a varied dc forward voltage pedestal, i.e. by shifting of Fermi level. Injection of minority carriers enables ones to eliminate the minority carrier traps (those seems to be efficient as the compensation centres within a lower half of a band-gap), and to regenerate barrier capacitance, as shown in Fig. 7.3. Enhancement of a forward dc voltage pedestal leads to filling of these traps. This determines an increase of the

60

BELIV current signal amplitude and a shift of the peak to the initial time instants within a transient, as illustrated in Fig. 7.3. Calibration of the injected density of minority carriers, when possible, and fixation of a reached saturation level for the amplitude of BELIV current transients can be exploited for evaluation of the density of the compensation centres. However, a suppression of the compensating centres at room temperature has been reached in our experiments only for the moderately irradiated ($\leq 10^{14}$ n/cm^2) diodes.

An impact of the radiation induced generation centres can also be suppressed by a reduction of temperature. Then, the carrier emission lifetime increases about exponentially with a reduction of temperature. The temperature dependent BELIV current transients are illustrated in Fig. 7.4a for the irradiated (10^{14} cm^{-2}) and reverse biased pin diode. It is clearly seen that the component of the carrier generation current decreases with reduction of temperature. Then, the shape of the BELIV current transient approaches to that inherent for dielectric capacitor, at 172 K. Thus, the heavily irradiated diode is fully depleted at equilibrium, although at room temperature this capacitor-inherent characteristic is masked by the large generation current component.

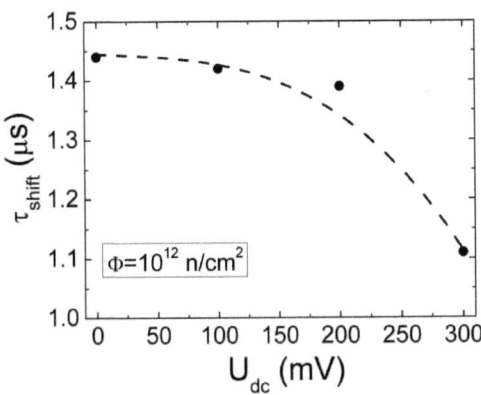

Figure 7.3. A shift of the barrier capacitance charging current peak position within transients recorded for reverse biased LIV dependent on forward dc voltage bias pedestal. (After E.Gaubas et al, ISRN Materials Science, (2012) article ID543790, doi:10.5402/2012/543790, [12]).

With reduction of temperature, which leads to a consequent increment of the carrier emission time and to a decrease of density of the empty capture-emission centres, the steady-state bias illumination becomes sufficient for suppression of the carrier capture centres. It can be seen in Fig. 7.4b, the barrier capacitance in moderately irradiated ($\leq 10^{14}$ n/cm^2) diodes restores to the values inherent for the lightly irradiated diodes when a combined conditioning made by the temperature lowering and additional illumination is applied. However, neither steady-state biasing by dc forward voltage pedestal nor continuous wave illumination is sufficient to suppress charge compensation and carrier capture/generation centres in heavily irradiated (> 10^{15} n/cm^2) diodes.

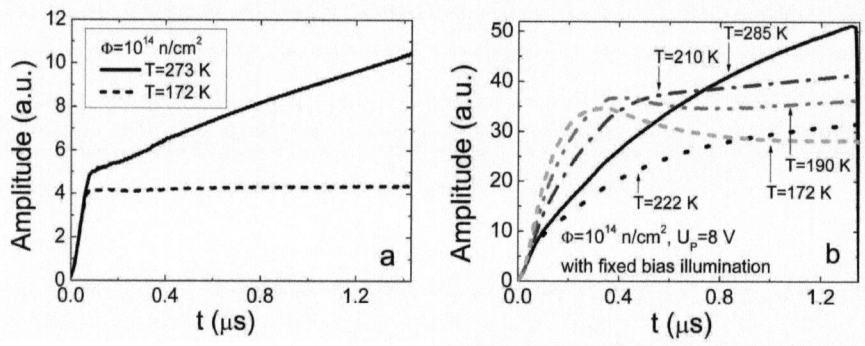

Figure 7.4. a- Temperature dependent variations of BELIV current transients in heavily (10^{14} n/cm^2) irradiated Si pin diode. b- Temperature dependent variations of the BELIV current transients recorded on the sample illuminated with light relevant to fill traps. (After E.Gaubas et al, ISRN Materials Science, (2012) article ID543790, doi:10.5402/2012/543790, [12]).

There are several options for evaluation of the diode parameters by using BELIV technique. The fluence dependent variations of the parameters of C_{b0}, U_{bi}, τ_g, N_{Def}, and N_T can be examined by using peculiar points and segments on the BELIV current transients, exploiting approximations expressed by equations Eqs. (2.2 - 2.10), if rather small voltages are applied. This enables ones to ignore the non-linear voltage sharing between R_L-C_{b0} elements of a circuit. The numerical simulations, using approximations described in equations Eqs. (2.2-2.10) and

$U_c(t)=U_P(t)-i_\Sigma(t)(R_L+R_{s//}(t))$, are inevitable if both the high precision of evaluation and a wide range of voltages are desirable. Here, $R_{s//}(t)$ represents a serial bulk resistance which can be approximated as $R_{s//}(t)=(d-w(t))/Se\mu n(t)$.

An illustration of the fitting procedure by numerical simulations, applied to the case of a diode irradiated with $\Phi=10^{12}$ cm^{-2}, is presented in Fig. 7.5a. To increase precision and to reduce an impact of the non-ideality of LIV pulses, a family of transients measured by varying LIV pulse ramp (A) have been simultaneously fitted with simulated ones (Fig. 7.5a). Here, extraction of parameters had been controlled by a non-linear-least-square (NLS) algorithm. A set of parameters of the U_{bi}, N_D and τ_g have been either kept invariable or simultaneously altered for all the transients within a family of curves. The LIV pulse parameters U_P and dU_P/dt had been taken from the experimental measurements within this fitting technique. The extracted parameters obtained from the best fit between the experimental and the simulated transients for NLS minimum are shown in the legend of Fig. 7.5a.

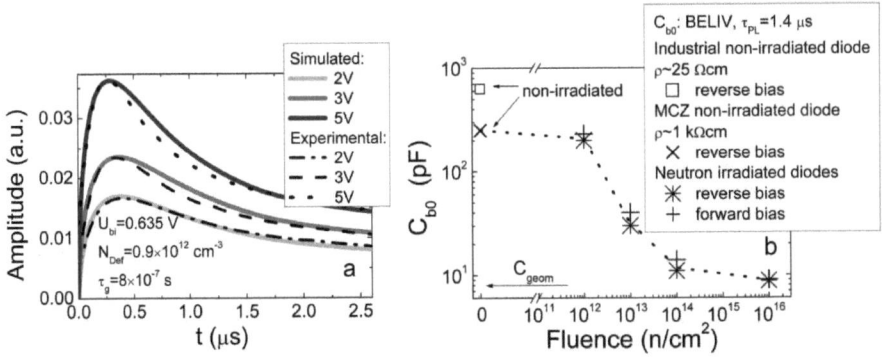

Figure 7.5. a - A family of transients, measured by varying LIV pulse ramp, fitted by simulated ones, when minimum for least-square deviations between simulated and experimental transients has been obtained. The set of parameters U_{bi}, N_{Def} and τ_g has been simultaneously extracted by this fitting procedure. The extracted values are denoted within a legend. b - Comparison of the fluence dependent variations of the barrier capacitance C_{b0} measured by BELIV technique employing both reverse (stars) and forward (crosses) LIV pulses (left scale) at T=300K. Values of C_{b0} obtained for non-irradiated Si diodes of different technology are also shown in Fig. 7.5b. (After E. Gaubas et al, ISRN Materials Science, (2012) article ID543790, doi:10.5402/2012/543790, [12]).

The extracted parameter C_{b0} as a function of irradiation fluence is shown in Fig. 7.5b. The initial barrier capacitance C_{b0} values, extracted by both the reverse and forward LIV pulses, coincide within measurement errors (those not exceed the size of symbols in Fig. 7.5b). The C_{b0} -Φ characteristic indicates a clear decrease of the absolute C_{b0} values due to outspreading of the depletion width. The decrease of C_{b0} is caused by an enhancement of w_0 due to a reduction of n_0.

Radiation induced defects are fast traps those reduce the effective density of dopants N_{Def} by formation of the compensating centres. Consequently, a barrier U_{bi} is modified. The de-trapped charge leads to an increase of the generation current component when trap density is significantly enhanced with fluence.

8. In situ barrier control during irradiation with hadrons

The on-line control of the production of radiation defects is an important issue within examination of the radiation damage of a detector. The BELIV technique appeared to be simply applicable for control of barrier and other device operational characteristics during exposure to high energy neutron and proton beams [35, 36].

The non-irradiated pin diodes of CERN standard p^+nn^+ detector structure of dimensions 5×5 mm^2 with doping density $N_D \cong 10^{12}$ cm^{-3} in n-Si 300 μm-thick layer [20] were employed for the on-line measurements. These particle detectors were fabricated from n- type MCZ (Czochralski with applied magnetic field grown) Si supplied by Okmetic Ltd. The samples were mounted on a relevant holder supported with the proper electrodes and signal transfer lines of about 10 m length for the remote measurements. The control instruments are placed within a safe area. Location of the sample within a centre of a particle beam has been aligned by using a laser collimation system. For the simultaneous fluence control, the Bruker alanine tablet has been mounted close to the detector sample. The spallator type neutrons source of about 25 MeV energy is based [37] on cyclotron accelerated deuterons beam interacting with Be target, which generates a cone of neutrons just behind a target. A neutrons flux can be easily manipulated by varying position of sample (on holder with necessary probes) relatively to a Be target. Also protons of 2-8 MeV energy accelerated by Tandetron-type accelerators have been employed for irradiation. For the latter measurements, sample has been mounted within a vacuum chamber.

An evolution of BELIV transients is illustrated in Fig. 8.1. Two main components in these transients represent the barrier capacitance charging current (i_C), the initial one, and the generation current (i_g), when observable increase in the rearward range of the BELIV pulses. These components [35] can be expressed, similarly to that derived Eqs. (2.2- 2.6), as

$$i(t) = i_C(t) + i_g(t) = \frac{\partial U}{\partial t}(C_b + U\frac{\partial C_b}{\partial w}\frac{\partial w(t)}{\partial U(t)}) + \sum_T \frac{em_{0,T}\,Sw(U(t))}{\tau_{g,T}} \quad .$$
(8.1)

The generation current is a sum over all types of traps T, distributed over $w(t)$, which represents thermally released carriers of density m_0, with a specific generation lifetime, commonly assumed as $\tau_g = N_{CS}\tau_V\tau_T exp(\Delta E_A/kT)$. The generation current (i_g) increases with $w(t)$.

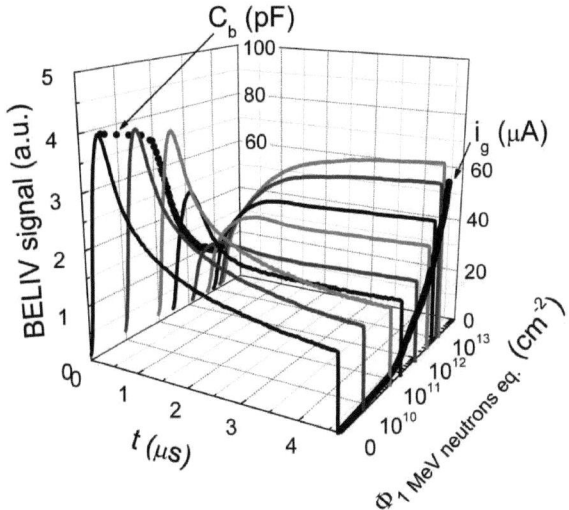

Figure 8.1. Variations of the BELIV transients as a function of 25 MeV neutron irradiation fluence, evaluated from exposure time of the in situ experiments. Variation of values of the barrier capacitance and generation current, extracted from these BELIV transients dependent on are additionally shown by dotted and solid curves, respectively.

These i_C and i_g current components compete and determine the observed changes of the shape of a BELIV pulse. The extracted values of barrier capacitance and of generation current as a function of fluence are shown on different planes of Fig. 8.1.

The clearly noticeable decrease of C_b and the enhancement of i_g correlate well mutually. It has been obtained that the decrease of C_b correlates with that of recombination lifetime reduction as a function of enhancement of the neutron irradiation fluence. The obtained reduction of $C_b \sim N_{Def}^{1/2}$ and the prevailing increase of $i_g \sim m_{0,T} \sim T$ (with T, as a density of specific type traps) indicate that radiation induced traps are the efficient recombination/generation centres. These traps compensate N_{Def} dopants leading to the degradation and further destruction of a junction.

The observed changes of the components of the barrier capacitance and of the generation current within BELIV transients correlate mutually when considered relative to an increasing fluence value. Approach of values of the carrier lifetime to that range of a specific time scale of charge drift leads to the non-operational junction. Thus, values of carrier recombination/generation lifetime determined from measured BELIV transients, can be employed for a prediction of the detector performance. It has been deduced that the employed type detectors are nearly destructed after irradiation with 4×10^{14} n/cm^2 1 MeV eq. fluence. The observed increase of generation current, even in the case of a rather short drift pulse, will cause considerable increase of the detector noise level.

The discussed illustrations of the on-line continuous scan of the changes of barrier capacitance and of generation current prove the applicability of the BELIV technique for the sensitive and remote probing of the evolution of radiation defects and device performance.

9. Profiling of junction location in solar-cell structures

Commercial solar-cells of a large area are fabricated using different technologies. These technologies are under permanent development. In creation of advanced technologies, where the rather thin layers of large conductivity are employed, the standard techniques, such as C-DLTS, TST LRC, for inspection of material quality are often non-applicable. Therefore, search of techniques for the comprehensive characterization of technological developments is relevant.

Application of the BELIV technique as a barrier control tool, combined with a step-positioning of the needle-tip probe located on the cross-sectional boundary of a parallel-plate layered structure has been tested and approved for the technological control of solar-cells. The designed experimental arrangement also enables ones to control the spreading resistance, carrier injection efficiency and the parameters associated with barrier capacitance of a junction.

9.1. Samples

Various n^+p solar-cell fragments of different top layer thickness and of junction area have been examined as the tentative crystalline Si solar-cell elements. The resistivity of a p-layer in solar-cells significantly exceeds that of high conductivity layer. The large conductivity and barrier capacitance values lead to the enhanced currents. This raises the specific requirements for sample preparation, - flatness of boundary cut, surface area of the sample, etc. Variations of the peak amplitudes have then been examined in order to resolve an interface location within a n^+p junction structure.

9.2. Profiling of junction location

In order to verify whether the BELIV technique is proper to resolve the rather thin layers, the solar-cell tentative n^+-p structures, containing 1- 4 μm thick n^+ and metallization layers have been tested.

Depth distribution of the amplitude scanned on the solar-cell structures are illustrated in Fig. 9.1. In the insets for Fig. 9.1, the inherent BELIV current transients, associated with different layers, are shown. Variations of the peak amplitudes enabled us to clearly resolve an interface location for a solar-cell structure.

Comparison of the profiles scanned on metallized and non-metallized solar cells (Fig. 9.1) enabled us to distinguish the top 4 μm-thick n^+ layer and the metallic electrode. Here, only fragments (0- 15 μm) of the scanned profiles nearby the junction interface are shown. The profiles are plotted starting from the surface of the structure. In the case of the metal electrode deposited on heavily doped layer, the transient repeats the LIV pulse shape over the length of a sum of the $d_{n+} + d_{metal\ electrode}$ thicknesses. Thickness of n^+ layer is consequently evaluated to be ~1 μm for a sample SC1 and ~4 μm for a sample SC-2, respectively. The 1.25 μm scanning step was kept in the latter profiling measurements.

The main errors in evaluation of layer thickness d_{n+} appear due to inhomogeneity of thickness of metallization layer. The reduction of an area of the main plate electrode, for the sample SC2, leads to a decrease of a barrier capacitance and, consequently, to a decrement of the BELIV current peak, relatively to SC1 sample. The outspread of the interfacial depth, with inherent increase of BELIV current amplitude (decrease of $R_{S//}$ and enhancement of the effective doping density), can be deduced within profiles, presented in Fig. 9.1, going towards bulk of the p-layer.

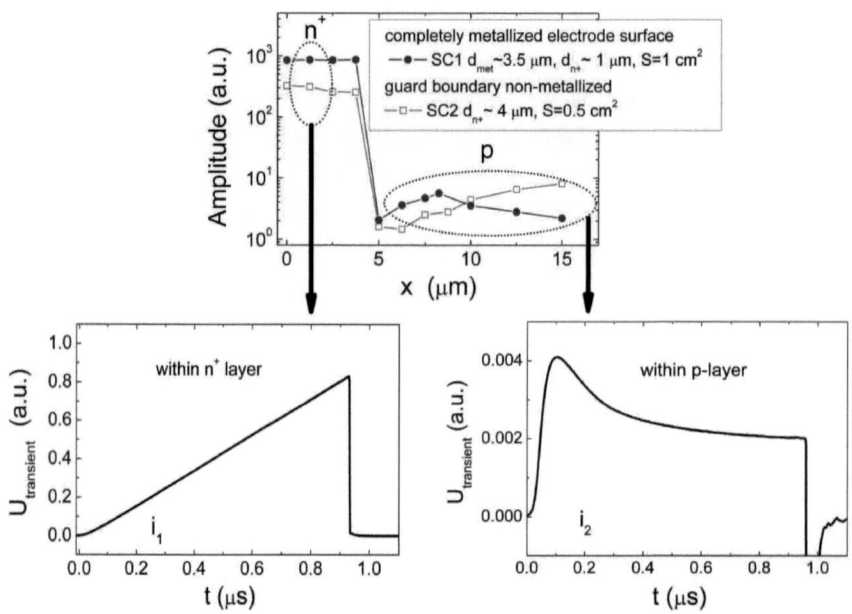

Figure 9.1. Profiles of the BELIV current amplitude in the solar-cell structure containing different (metallization/n^+) thickness and area of electrode. The shapes of the registered current transients within respective depth are shown in the insets. (After E.Gaubas et al, ISRN Materials Science, (2012) article ID543790, doi:10.5402/2012/543790, [12]).

Thus, the BELIV profiling technique is even suitable for scans on the rather thin layered structures using a rather wide ($a \approx 10$ µm) probe. As can be noticed in Figs. 3.1a and 3.1b for simulated potential and spread of currents, the enhanced potential and current density appear at the front edge of the probe. Therefore, it is confirmed that resolution of scanning can be significantly better than a width (a) of the needle-tip probe.

Examination of a rearward component in the illustrated BELIV transient (inset of Fig. 9.1) revealed an impact of the space charge generation current in the solar-cell structures. This has been highlighted by increasing a LIV pulse duration to match the time scale for carrier thermal emission.

The significant impact of carrier generation processes, in time scale exceeding 10 μs, has been corroborated by the DLTS measurements on the same structures of Si solar cells, where traps ascribed to the metal impurities have been identified.

10. Thermal emission spectroscopy in the industrial solar-cells

The efficiency of a solar cell significantly depends on the technological defects introduced during the formation of junctions, passivation layers and electrodes [38, 39]. The alternative technology development is a solar cell production, using copper electrodes. Such a technology is attractive, in order to cheapen the commercial fabrication of solar cells. However, the problem of copper diffusion into the base region is inevitable. This problem can partially be solved by forming the diffusion barrier, which prevents copper in-diffusion [38-40]. The formation of copper electrodes on a solar cell is additionally complicated by the necessity to combine several processes of a deposition of different metals. These technological procedures should be precisely controlled to avoid a contamination of the active device layers.

The problem of the usage of the standard techniques and instrumentation (such as the capacitance deep level transient spectroscopy [6, 7], C-DLTS) then appears in precise control of the introduced defects within a high conductivity Si material. The high conductivity base region leads to the large barrier capacitance charging currents. Additionally, the enhanced leakage current on sample boundaries is inevitable for the small area samples, required for the application of the standard C-DLTS instruments and methods. Therefore, it is important to develop the alternative techniques for the defect spectroscopy in the high conductivity junction structures, in order to directly control a low density of metallic impurities within the active layers of solar-cells. Therefore, the BELIV based spectroscopy of the technological defects have been developed by combining the temperature scans of the thermal generation currents extracted from barrier capacitance charging transients and the capacitance deep level transient spectroscopy (C-DLTS) techniques.

10.1. Samples and experimental arrangements

The tentative industry solar cells made employing of the copper technology were examined. A solar cell contains the 180 μm thick p-Si base boron doped region. The n^+-p junction with an emitter thickness of 0.3 μm was formed. The anti-reflection and the 150 nm thick Si_3N_4 surface passivation layers were formed. The silicide layers of Ni_2Si of thickness of 0.1 μm were also deposited. The electrodes were formed depositing by chemical means the Ni and Cu layers of 250 nm and 20 μm thickness, respectively.

The solar cell fragments of 4×4 mm^2 were cut for the C-DLTS measurements, as the commercial DLTS instruments enable compensation of the steady-state barrier capacitance of values less than 3 nF. It has been found within the DLTS measurements that the nickel-copper electrodes are insufficiently proof for the temperature cycling in the range of 50-300 K. Therefore, several samples were made using the 100 nm-thick gold electrodes deposited by a magnetron sputtering, after the copper electrode had been removed.

The C-DLTS measurements have been implemented by using a HERA-DLTS System 1030 spectrometer. The sample is connected to electrodes on sample holder within a closed-cycle He cryostat. The barrier capacitance changes (due to carrier traps) are directly recorded as the capacitance transients by using a Boonton capacitance meter installed within the HERA-DLTS System 1030 spectrometer. The capacitance transient signals are transferred to the base unit of the spectrometer, which is equipped with a voltage source, a low-pass filter and an amplifier. The C-DLTS measurement procedures are controlled by a personal computer with installed Phys-Tech software. This Phys-Tech software also contains a trap identification library using the extracted trap signatures such as the carrier capture cross-section and trap thermal activation energy. To avoid the copper electrode damage caused by temperature cycling, the frequency scans at fixed temperatures have been made for the primary defect identification.

To clarify the impact of the Ni/Cu electrode thermal instability and to exclude the leakage current enhancement (inherent for the small area junction structures), the BELIV temperature scans were performed on the bigger fragments of the industrial solar cell using the Ni/Cu and Au top-electrodes.

10.2. Spectroscopy of technological defects

In Fig. 10.1, the temperature dependent changes of the BELIV transients are illustrated for both the Ni/Cu (a) and Au (b) top-electrodes.

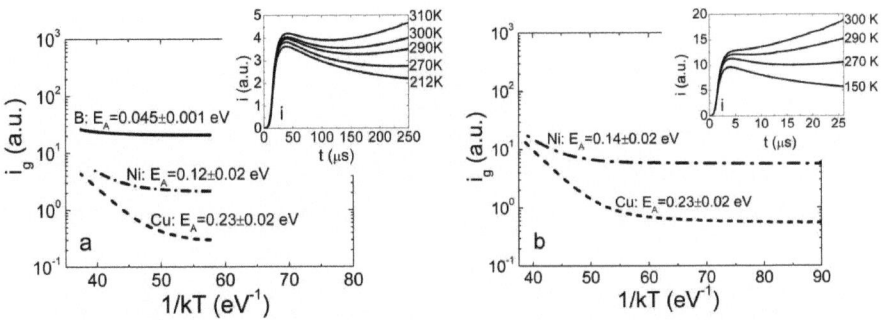

Figure 10.1. (a) Generation current extracted from the BELIV transients recorded at 26 μs (top curve), 260 μs (middle curve) and 2.6 ms (bottom curve) LIV pulse duration, respectively, on solar cell fragment containing Ni/Cu electrodes. (b) Generation currents measured on solar cell fragment containing the Au top electrode with LIV pulse durations of 26 μs (top curve) and 260 μs (bottom curve), respectively. Temperature dependent variations of the BELIV transients are illustrated in the insets.

It can be noticed in Fig. 10.1. that the resolvable generation current increase within the recorded BELIV transients is obtained for temperatures above 270 K. Also, the current is larger for the gold top-electrode containing samples. While, the generation current can reliably be resolved for the longer LIV pulse duration in Ni/Cu electrode containing solar-cell fragments. However, the measurements using Ni/Cu electrode can be performed only a few times, due to the destroyed adhesion of the metal layer. As discussed in presentation of the BELIV method principles, the generation current is added to the barrier capacitance charging/charge extraction current component in BELIV transients. Therefore, the sum of the currents increases with enhancement of the thermal emission from traps. On the other hand, the increase of the LIV pulse duration leads to a reduction of the LIV pulse ramp A, and this yields a decrease of the BELIV current signal. Therefore, the decrease of the BELIV signal with enhancement of LIV pulse duration is a disadvantage of the BELIV technique, in comparison with C-DLTS, for detection of very deep levels. However, C-DLTS is insufficiently sensitive to traps with short thermal emission lifetime. Thereby, combining of these C-DLTS and BELIV techniques can be efficient tool for examination of the large conductivity and capacitance samples.

The BELIV transients have been recorded by varying durations of LIV pulses in the range from 2.6 μs to 2.6 ms. The generation current component is extracted from the BELIV transients, and its dependence on temperature has then been examined. The different levels are temperature scanned for the different fixed LIV pulse durations: the shallower the examined level is, the shorter LIV pulse should be employed. Thereby, several temperature scan curves are obtained by varying the LIV pulse duration. These curves then show different activation energy values, denoted in Figs. 10.1a and 10.1b. The activation energy is represented by a slope of generation current plotted in the semilog scale as a function of the reverse thermal energy kT.

The traps with activation energy values of E_A=0.045±0.001 eV, 0.12±0.02 eV and 0.23±0.02 eV have been resolved in solar cell fragments containing Ni/Cu electrodes. These traps can be assigned to boron dopants with E_A=0.045±0.001 eV, and to the nickel impurities with E_A=0.12±0.02 eV. The trap with activation energy of E_A=0.23±0.02 eV can be ascribed to copper impurities. This latter activation energy value might also be inherent to Ni, as referenced in literature [41], however, the value of the capture cross-section, attributed to Cu, is considerably larger. Therefore, the trap with activation energy of E_A=0.23±0.02 eV is assigned to the copper impurities in this consideration. In samples made of solar cell fragment containing Au top electrode, the Ni ascribed traps with E_A=0.12±0.02 eV and the Cu attributed ones with activation energy of E_A=0.23±0.02 eV have been resolved, as well.

The C-DLTS measurements corroborated existence of several traps with DLTS signatures, namely: E_A=0.41±0.01 eV, T_{peak}~242 K and E_A=0.22±0.01 eV, T_{peak}~120 K with capture cross-section σ~5×10^{-13} cm^2 are inherent for the Cu ascribed traps in Si [41]; the activation energy of E_A~0.21- 0.23 eV, is attributed to the Ni impurities in Si, as referenced in literature [41], although with the less value (of σ~10^{-16} cm^2) of the carrier capture cross-section.

This indicates that Ni and Cu impurities were incorporated during the primary formation of electrodes on solar cells. The Cu impurities seem to form several deep levels with thermal activation energy values of E_A=0.23±0.02 eV, as determined from BELIV and C-DLTS temperature scans, and E_A=0.41±0.02 eV, as extracted from C-DLTS spectroscopy data.

Summary

A new pulsed technique of barrier evaluation by linearly increasing voltage (BELIV) for inspection of both barrier and diffusion capacitances in junction structures has been developed and approved for the fast and comprehensive characterization of junction structures [12, 13, 16, 19, 25, 26, 35, and 36]. Different aspects of this technique are reviewed in this monograph. The principles of parameter extraction by using BELIV technique are described in detail for the reverse and forward bias regimes [11, 13, 16, and 19]. It has been proved that the measured transients represent an evolution of the charging currents after the surface charge discontinuity is stabilized at the interface, located at depletion width. The Ramo's current components of a few ns durations, ascribed to the surface charge discontinuity drift, can only appear as a noise signal on the slower signal changes ascribed to the barrier capacitance charging currents, for LIV pulses of microsecond scale durations. Therefore, BELIV transients can be analyzed using the depletion approximation.

It has been demonstrated that BELIV technique is a tool for the rapid evaluation of junction parameters [12, 19]. This technique has been approved on different structures, and it is suitable for examination of the structures of irradiated particle detectors, pin-diodes, thyristors, solar–cells, and hetero-junctions in polycrystalline materials [12, 13, 16, 19, 25, 26, 35, and 36]. Variations of parameters of the recombination τ_R and generation τ_g lifetimes as a function of irradiation fluence are in good agreement with those measured by microwave probed photoconductivity transient technique on the same samples. Temperature dependent variations of the $\tau_g(T)$ and $\tau_R(T)$ parameters can be applied for spectroscopy of deep levels, when the current components ascribed to τ_g and τ_R are distinguishable within BELIV current transients [25]. It has been unveiled that the generation current component prevails in heavily irradiated diodes [25, 35].

The BELIV pulsed technique enables ones to clarify a few significant aspects: to identify charge extraction regime and to estimate barrier capacitance, to clarify competition between the barrier capacitance charging (i_C) and generation (i_g) currents, to clarify the full depletion state for heavily irradiated diodes [25]. It has been shown [25] that the built-in full depletion (insulating state) is inherent for Si diodes doped with 10^{12} cm^{-3} donors density when hadron irradiation fluences exceed 10^{14} cm^{-2}.

The presented BELIV technique is applicable for fast estimation of actual location of barriers [12. 13], of doping profiles [13] and of deep traps within layered junction structures [16]. This technique has been applied and appeared to be suitable for depth-scans of the pin-diode, thyristor and solar–cell structures [12, 13]. Depth - variations of the amplitude and duration of the BELIV current transients and of the shape of these transients are in good agreement with such characteristics evaluated by using common spreading-resistance profiling instruments [13]. The parameters of barrier capacitance, of density of the effective doping, of space charge generation current, extracted from the characteristics of BELIV current transients can be applied for spectral analysis of deep levels [16]. The spectral signatures obtained by combining the photo-ionization and thermal emission spectroscopy implemented using BELIV technique are in excellent agreement with those determined by DLTS techniques on the same samples [12, 16]. BELIV technique can be employed as a fast tool to obtain indications on existence of deep centres within definite layers of heterostructures [28] for more detail identification of these traps by using the DLTS spectroscopy [42].

The combined study by using the capacitance deep level transient spectroscopy and the generation current temperature scan techniques enables ones to resolve the metal impurities. This can be applied in development of new technologies in fabrication of industrial solar cells and to control the technological procedures in formation of solar cells [12].

It has been proved the applicability of the BELIV technique for the sensitive and remote probing of the evolution of radiation defects and device performance during high energy neutron and proton irradiations [35, 36].

References

1. K.Blotekjer, *Transport equations for electrons in two-valley semiconductors*, IEEE Trans. Electr. Dev., **ED-17** (1970) 38.
2. B.J.Baliga, *Power semiconductor devices*, PWS Publishing Company, Boston, 1995.
3. W. van Roosbroeck, *The transport of added current carriers in a homogeneous semiconductor*, Phys. Rev., **91** (1953) 282.
4. A.Shockley, *Currents to conductors induced by a moving point charge*, J. Appl. Phys., **9** (1938) 635.
5. E.Gaubas, T.Ceponis, and V.Kalesinskas, *Currents induced by injected charge in junction detectors*, Sensors **13** (2013) 12295.
6. D.K.Schroder, *Semiconductor material and device characterization*, 3rd Ed., John Wiley and Sons, New Jersey, 2006.
7. P.Blood and J.W.Orton, *The electrical characterization of semiconductors: majority carriers and electron state*, Academic Press, New York, 1992.
8. S.Ramo, *Currents induced by electron motion*, Proc. I.R.E. **27** (1939) 584.
9. S.M.Sze, Kwork K.Ng, *Physics of semiconductor devices*, (John Wiley & Sons, NJ, 2007).
10. S. M. Sze, *Physics of semiconductor devices*, John Wiley and Sons, New York, 1981.
11. E.Gaubas, T.Ceponis, J.Vaitkus, and J.Raisanen, *Study of variations of the carrier recombination and charge transport parameters during proton irradiation of silicon pin diode structures*, AIP Advances, **1** (2011) 022143.
12. E.Gaubas, T.Čeponis, V.Kalendra, J.Kusakovskij, and A.Uleckas, *Barrier evaluation by linearly increasing voltage technique applied to Si solar cells and irradiated pin diodes*, ISRN Materials Science, (2012) article ID543790, doi:10.5402/2011/543790.
13. E.Gaubas, T.Ceponis, and J.Kusakovskij, *Profiling of barrier capacitance and spreading resistance by transient linearly increasing voltage technique*, Review of Scientific Instruments, **82** (2011) 083304.
14. G.Lucovsky, *On the photoionization of deep impurity centers in semiconductors*, Solid State Commun., **3** (1965) 299.
15. E.Gaubas, A.Uleckas, R.Grigonis, V.Sirutkaitis, and J.Vanhellemont, *Microwave probed photoconductivity spectroscopy of deep levels in Ni doped Ge*. Appl. Phys. Lett., **92** (2008) 222102.

16. E.Gaubas, T.Ceponis, A.Uleckas, and R.Grigonis, *Room temperature spectroscopy of deep levels in junction structures using barrier capacitance charging current transients*, Journal of INSTrumentation, **7** (2012) P01003, doi:10.1088/1748-0221/7/01/P01003.

17. A.Chantre, G.Vincent, and D.Bois. *Deep level optical spectroscopy in GaAs.* Phys. Rev. B, **23** (1981) 5335.

18. J.Bourgoin and M.Lanoo, *Point defects in semiconductors* II *Experimental aspects,* Springer-Verlag, Berlin, 1983.

19. E.Gaubas, T.Čeponis, S.Sakalauskas, A.Uleckas, and A. Velička, *Fluence dependent variations of barrier and generation currents in neutron and proton irradiated Si particle detectors*, Lithuanian Journal Physics, **51** (2011) 227.

20. www.cern.ch/rd50.

21. H.G.Grimmeiss, *Deep level impurities in semiconductors*, Annu. Rev. Mat. Sci., **7** (1977) 341.

22. J.W.Chen and A.G.Milnes, *Energy levels in silicon*, Annu.Rev. Mat. Sci., **10** (1980) 157.

23. M.Karimov and A.Karakhodzhaev, *Influence of the post-diffusion hardening rate and thermal treatment on the thermal stability of the charge-carrier lifetime in overcompensated n-Si(B,S)*, Russian Physics Journal, **44** (2001) 734.

24. M.Huhtinen, *Simulation of non ionising energy loss and defect formation in silicon*, Nucl. Instrum. and Meth. in Phys. Res., **A 491** (2002) 194.

25. E.Gaubas, T.Ceponis, and J.Vaitkus, *Impact of generation current on the evaluation of the depletion width in heavily irradiated Si detectors,* J. Appl. Phys. **110** (2011) 033719.

26. T.S.Te Velde, *The production of the cadmium sulphide-copper sulphide solar cells*, Energy Conv., **14** (1975) 111.

27. A.Goldenblum, A.Oprea, *Photocapacitance effects in dry processed Cu_2S-CdS heterojunctions*, J. Phys. D: Appl. Phys. **27** (1994) 582.

28. E.Gaubas, I.Brytavskyi, T.Čeponis, J.Kusakovskij, and G.Tamulaitis, *Barrier capacitance characteristics of CdS-Cu_2S junction structures*, Thin Solid Films, **531** (2012) 131, doi:10.1016/j.tsf.2013.01.010.

29. P.F.Ermolov, D.E.Karmanov, A.K.Leflat, V.M.Manankov, M.M.Merkin and E.K.Shabalina, *Neutron-irradiation-induced effects caused by divacancy clusters with a tetravacancy core in float-zone silicon*, Semiconductors, **36** (2002) 1114.

30. J.C.Vickerman and I.S.Gilmore, *Surface analysis – the principal techniques,* 2nd Ed., John Wiley and Sons, Chichester, 2009.

31. S.Kalinin and A.Gruverman, *Scanning Probe Microscopy*, vol.1, Springer, New York, 2007.
32. X.Ou, P.Das Kanungo, R.Kogler, P.Werner, U.Gosele, W.Skorupa, and X.Wang, *Carrier profiling of individual Si nanowires by scanning spreading resistance microscopy*, Nano Letters, **10** (2010) 171.
33. D.V.Lang, *Deep-level transient spectroscopy: a new method to characterize traps in semiconductors*, J.Appl. Phys., **45** (1974) 3023.
34. S.Vayrynen, J.Raisanen, I.Kassamakov, E. Tuominen, *Breakdown of silicon particle detectors under proton irradiation*, J. Appl. Phys., **106** (2009) 104914.
35. E.Gaubas, T. Ceponis, A.Jasiunas, A.Uleckas, J. Vaitkus, E.Cortina, O.Militaru, *Correlated evolution of barrier capacitance charging, generation and drift currents and of carrier lifetime in Si structures during 25 MeV neutron irradiation*, Appl. Phys. Lett., **101** (2012) 232104.
36. T. Ceponis, E. Gaubas, V. Kalendra, A. Uleckas, J. Vaitkus, K. Zilinskas, V. Kovalevskij, M. Gaspariunas and V. Remeikis, *In situ analysis of carrier lifetime and barrier capacitance variations in silicon during 1.5 MeV protons implantation*, J. Inst., **6** (2011) P09002.
37. K. Bernier, H. Boukhal, J.-M. Denis, T. El Bardouni, Gh. Grégoire, O. Grégoire, and V. Tran, *An intense fast neutron beam in Louvain- la- Neuve*, preprint at Nuclear Physics Institute, Université Catholique de Louvain, Louvain-la-Neuve, Belgium,- www.cyc.ucl.ac.be.
38. J.S.You, J.Kang, D.Kim, J.J.Pak, and C.S.Kang, *Copper metallization for crystalline Si solar cells*, Solar Energy Materials and Solar Cells **79** (2003) 339.
39. K.M.Chow, W.Y.Ng, and L.K.Yeung, *Barrier properties of Ni, Pd and Pd-Fe for Cu diffusion*, Surface and Coating Technology, **105** (1998) 56.
40. L.C.Leu, D.P.Norton, L.McElwee-White, and T.J.Anderson, *Ir/TaN as a bilayer diffusion barries for advanced Cu interconnects*, Appl. Phys. Lett., **92** (2008) 111917.
41. K.Graff, *Metal impurities in silicon-device fabrication*, 2nd ed., Springer, 2000.
42. E.Gaubas, I.Brytavskyi, T.Ceponis, V.Kalendra, and A.Tekorius, *Spectroscopy of deep traps in Cu_2S-CdS junction structures,* Materials, **5** (2012) 2597.

Printed by Books on Demand GmbH, Norderstedt / Germany